CORVETTE STING RAY
Gold Portfolio
1963~1967

Compiled by
R.M. Clarke

ISBN 1 85520 0244

Distributed by
Brooklands Book Distribution Ltd.
'Holmerise', Seven Hills Road,
Cobham, Surrey, England

Printed in Hong Kong

BROOKLANDS BOOKS

BROOKLANDS ROAD TEST SERIES
AC Ace & Aceca 1953-1983
Alfa Romeo Alfasud 1972-1984
Alfa Romeo Alfetta Coupes GT. GTV. GTV6 1974-1987
Alfa Romeo Giulia Berlinas 1962-1976
Alfa Romeo Giulia Coupes 1963-1976
Alfa Romeo Spider 1966-1987
Allard Gold Portfolio 1937-1958
Alvis Gold Portfolio 1919-1969
American Motors Muscle Cars 1966-1970
Aston Martin Gold Portfolio 1972-1985
Austin Seven 1922-1982
Austin A30 & A35 1951-1962
Austin Healey 3000 1959-1967
Austin Healey 100 & 3000 Col No.1
Austin Healey 'Frogeye' Sprite Col No.1 1958-1961
Austin Healey Sprite 1958-1971
Avanti 1962-1983
BMW Six Cylinder Coupes 1969-1975
BMW 1600 Col. 1 1966-1981
BMW 2002 1968-1976
Bristol Cars Gold Portfolio 1946-1985
Buick Automobiles 1947-1960
Buick Muscle Cars 1965-1970
Buick Riviera 1963-1978
Cadillac Automobiles 1949-1959
Cadillac Automobiles 1960-1969
Cadillac Eldorado 1967-1978
High Performance Capris Gold Portfolio 1969-1987
Chevrolet Camaro & Z-28 1973-1981
High Performance Camaros 1982-1988
Chevrolet Camaro Col No.1 1967-1973
Camaro Muscle Cars 1966-1972
Chevrolet 1955-1957
Chevrolet Impala & SS 1958-1971
Chevrolet Muscle Cars 1966-1971
Chevelle and SS 1964-1972
Chevy EL Camino & SS 1959-1987
Chevy II Nova & SS 1962-1973
Chrysler 300 1955-1970
Citroen Traction Avant Gold Portfolio 1934-1957
Citroen DS & ID 1955-1975
Citroen 2CV 1949-1988
Shelby Cobra Gold Portfolio 1962-1969
Cobras & Replicas 1962-1983
Corvair 1959-1968
Chevrolet Corvette Gold Portfolio 1953 1962
Corvette Stingray Gold Portfolio 1963-1967
High Performance Corvettes 1983-1989
Datsun 240Z 1970-1973
Datsun 280Z & ZX 1975-1983
De Tomaso Collection No.1 1962-1981
Dodge Charger 1966-1974
Dodge Muscle Cars 1967-1970
Excalibur Collection No.1 1952-1981
Ferrari Cars 1946-1956
Ferrari Cars 1973-1977
Ferrari Dino 1965-1974
Ferrari Dino 308 1974-1979
Ferrari 308 & Mondial 1980-1984
Ferrair Collection No.1 1960-1970
Fiat-Bertone X1/9 1973-1988
Fiat Pininfarina 124 + 2000 Spider 1968-1985
Ford Automobiles 1949-1959
Ford GT40 Gold Portfolio 1964-1987
Ford Fairlane 1955-1970
Ford Falcon 1960-1970
High Perfomance Mustangs 1982-1988
Ford Cortina 1600E & GT 1967-1970
Ford RS Escorts 1968-1980
High Performance Escorts Mk1 1968-1974
High Performance Escorts Mk II 1975-1980
Honda CRX 1983-1987
Hudson & Railton 1936-1940
Jaguar Cars 1957-1961
Jaguar Cars 1961-1964
Jaguar Mk2 1959-1969
Jaguar E-Type Gold Portfolio 1961-1971
Jaguar E-Type 1966-1971
Jaguar E-Type V-12 1971-1975
Jaguar XKE Collection No.1 1961-1974
Jaguar XJ6 1968-1972
Jaguar XJ6 Series II 1973-1979
Jaguar XJ6 & XJ12 Series III 1979-1985
Jaguar XJ12 1972-1980
Jaguar XJS Gold Portfolio 1975-1988
Jaguar XK120.XK140.XK150 Gold Portfolio 1948-1960
Jeep CJ5 & CJ6 1960-1976
Jeep CJ5 & CJ7 1976-1986
Jensen Cars 1946-1967
Jensen Cars 1967-1979
Jensen Interceptor Gold Portfolio 1966-1986
Jensen Healey 1972-1976
Lamborghini Cars 1964-1970
Lamborghini Cars 1970-1975
Lamborghini Countach Col No.1 1971-1982
Lamborghini Countach & Urraco 1974-1980
Lamborghini Countach & Jalpa 1980-1985
Lancia Stratos 1972-1985
Land Rover 1948-1973 - A Collection
Land Rover Series II & IIa 1958-1971
Land Rover Series III 1971-1985
Land Rover 90 & 110 1983-1989
Lincoln Gold Portfolio 1949-1960
Lincoln Continental 1961-1969
Lotus and Caterham Seven Gold Portfolio 1957-1989
Lotus Elan Gold Portfolio 1962-1974
Lotus Elan Collection No.2 1963-1972
Lotus Elite 1957-1964
Lotus Elite & Eclat 1974-1982
Lotus Turbo Esprit 1980-1986
Lotus Europa 1966-1975
Lotus Europa Collection No.1 1966-1974
Lotus Seven Colleciton No.1 1957-1982
Marcos Cars 1960-1988
Maserati 1965-1970
Maserati 1970-1975
Mazda RX-7 Collection No.1 1978-1981
Mercedes 190 & 300SL 1954-1963

Mercedes 230/250/280SL 1963-1971
Mercedes Benz SLs & SLCs Gold Portfolio 1971-1989
Mercedes Benz Cars 1949-1954
Mercedes Benz Cars 1954-1957
Mercedes Benz Cars 1957-1961
Mercedes Benz Compeition Cars 1950-1957
Mercury Muscle Cars 1966-1971
Metropolitan 1954-1962
MG TC 1945-1949
MG TD 1949-1953
MG TF 1953-1955
MG Cars 1959-1962
MGA Roadsters 1955-1962
MGA Collection No.1 1955-1962
MGB Roadsters 1962-1980
MGB GT 1965-1980
MG Midget 1961-1980
Mini Moke 1964-1989
Mini Muscle Cars 1961-1979
Mopar Muscle Cars 1964-1967
Mopar Muscle Cars 1968-1971
Morgan Three-Wheeler Gold Portfolio 1910-1952
Morgan Cars 1960-1970
Morgan Cars Gold Portfolio 1968-1989
Morris Minor Collection No.1
Mustang Muscle Cars 1967-1971
Oldsmobile Automobiles 1955-1963
Old's Cutlass & 4-4-2 1964-1972
Oldsmobile Muscle Cars 1964-1971
Oldsmobile Toronado 1966-1978
Opel GT 1968-1973
Packard Gold Portfolio 1946-1958
Pantera Gold Portfolio 1970-1989
Plymouth Barracuda 1964-1974
Plymouth Muscle Cars 1966-1971
Pontiac Tempest & GTO 1961-1965
Pontiac GTO 1964-1970
Pontiac Firebird 1967-1973
Pontiac Firebird and Trans-Am 1973-1981
High Performance Firebirds 1982-1988
Pontiac Fiero 1984-1988
Pontiac Muscle Cars 1966-1972
Porsche 356 1952-1965
Porsche Cars in the 60's
Porsche Cars 1960-1964
Porsche Cars 1964-1968
Porsche Cars 1968-1972
Porsche Cars 1972-1975
Porsche Turbo Collection No.1 1975-1980
Porsche 911 1965-1969
Porsche 911 1970-1972
Porsche 911 1973-1977
Porsche 911 Carrera 1973-1977
Porsche 911 Turbo 1975-1984
Porsche 911 SC 1978-1983
Porsche 914 Gold Portfolio 1969-1976
Porsche 914 Collection No.1 1969-1983
Porsche 924 Gold Portfolio 1975-1988
Porsche 928 1977-1989
Porsche 944 1981-1985
Range Rover Gold Portfolio 1970-1988
Reliant Scimitar 1964-1986
Riley 11/2 & 21/2 Litre Gold Portfolio 1945-1955
Rolls Royce Silver Cloud 1955-1965
Rolls Royce Silver Shadow 1965-1981
Rover P4 1949-1959
Rover P4 1955-1964
Rover 3 & 3.5 Litre 1958-1973
Rover 2000 + 2200 1963-1977
Rover 3500 1968-1977
Rover 3500 & Vitesse 1976-1986
Saab Sonett Collection No.1 1966-1974
Saab Turbo 1976-1983
Shelby Mustang Muscle Cars 1965-1970
Studebaker Gold Portfolio 1947-1966
Stubebaker Hawks & Larks 1956-1963
Sunbeam Tiger & Alpine Gold Portfolio 1959-1967
Thunderbird 1955-1957
Thunderbird 1958-1963
Thunderbird 1964-1976
Toyota MR2 1984-1988
Triumph 2000. 2.5. 2500 1963-1977
Triumph GT6 1966-1974
Triumph Spitfire 1962-1980
Triumph Spitfire Col No.1 1962-1982
Triumph Stag 1970-1980
Triumph Stag Collection No.1 1970-1984
Triumph TR2 & TR3 1952-60
Triumph TR4-TR5-TR250 1961-1968
Triumph TR6 1969-1976
Triumph TR6 Collection No.1 1969-1983
Triumph TR7 & TR8 1975-1982
Triumph Vitesse & Herald 1959-1971
TVR Gold Portfolio 1959-1988
Volkswagen Cars 1936-1956
VW Beetle Collection No.1 1970-1982
VW Golf GTi 1976-1986
VW Karmann Ghia 1955-1982
VW Kubelwagen 1940-1975
VW Scirocco 1974-1981
VW Bus. Camper. Van 1954-1967
VW Bus. Camper. Van 1968-1979
VW Bus. Camper. Van 1979-1989
Volvo 120 1956-1970
Volvo 1800 1960-1973

BROOKLANDS ROAD & TRACK SERIES
Road & Track on Alfa Romeo 1949-1963
Road & Track on Alfa Romeo 1964-1970
Road & Track on Alfa Romeo 1971-1976
Road & Track on Alfa Romeo 1977-1989
Road & Track on Aston Martin 1962-1984
Road & Track on Auburn Cord and Duesenburg 1952-1984
Road & Track on Audi & Auto Union 1952-1980
Road & Track on Audi 1980-1986
Road & Track on Austin Healey 1953-1970
Road & Track on BMW Cars 1966-1974
Road & Track on BMW Cars 1975-1978
Road & Track on BMW Cars 1979-1983

Road & Track on Cobra, Shelby & GT40 1962-1983
Road & Track on Corvette 1953-1967
Road & Track on Corvette 1968-1982
Road & Track on Corvette 1982-1986
Road & Track on Datsun Z 1970-1983
Road & Track on Ferrari 1950-1968
Road & Track on Ferrari 1968-1974
Road & Track on Ferrari 1975-1981
Road & Track on Ferrari 1981-1984
Road & Track on Fiat Sports Cars 1968-1987
Road & Track on Jaguar 1950-1960
Road & Track on Jaguar 1961-1968
Road & Track on Jaguar 1968-1974
Road & Track on Jaguar 1974-1982
Road & Track on Jaguar 1983-1989
Road & Track on Lamborghini 1964-1985
Road & Track on Lotus 1972-1981
Road & Track on Maserati 1952-1974
Road & Track on Maserati 1975-1983
Road & Track on Mazda RX7 1978-1986
Road & Track on Mercedes 1952-1962
Road & Track on Mercedes 1963-1970
Road & Track on Mercedes 1971-1979
Road & Track on Mercedes 1980-1987
Road & Track on MG Sports Cars 1949-1961
Road & Track on MG Sprots Cars 1962-1980
Road & Track on Mustang 1964-1977
Road & Track on Peugeot 1955-1986
Road & Track on Pontiac 1960-1983
Road & Track on Porsche 1961-1967
Road & Track on Porsche 1968-1971
Road & Track on Porsche 1972-1975
Road & Track on Porsche 1975-1978
Road & Track on Porsche 1979-1982
Road & Track on Porsche 1982-1985
Road & Track on Porsche 1985-1988
Road & Track on Rolls Royce & B'ley 1950-1965
Road & Track on Rolls Royce & B'ley 1966-1984
Road & Track on Saab 1955-1985
Road & Track on Toyota Sports & GT Cars 1966-1984
Road & Track on Triumph Sports Cars 1953-1967
Road & Track on Triumph Sports Cars 1967-1974
Road & Track on Triumph Sports Cars 1974-1982
Road & Track on Volkswagen 1951-1968
Road & Track on Volkswagen 1968-1978
Road & Track on Volkswagen 1978-1985
Road & Track on Volvo 1957-1974
Road & Track on Volvo 1975-1985
Road & Track - Henry Manney at Large and Abroad

BROOKLANDS CAR AND DRIVER SERIES
Car and Driver on BMW 1955-1977
Car and Driver on BMW 1977-1985
Car and Driver on Cobra, Shelby & Ford GT 40 1963-1984
Car and Driver on Corvette 1956-1967
Car and Driver on Corvette 1968-1977
Car and Driver on Corvette 1978-1982
Car and Driver on Corvette 1983-1988
Car and Driver on Datsun Z 1600 & 2000 1966-1984
Car and Driver on Ferrari 1955-1962
Car and Driver on Ferrari 1963-1975
Car and Driver on Ferrari 1976-1983
Car and Driver on Mopar 1956-1967
Car and Driver on Mopar 1968-1975
Car and Driver on Mustang 1964-1972
Car and Driver on Pontiac 1961-1975
Car and Driver on Porsche 1955-1962
Car and Driver on Porsche 1963-1970
Car and Driver on Porsche 1970-1976
Car and Driver on Porsche 1977-1981
Car and Driver on Porsche 1982-1986
Car and Driver on Saab 1956-1985
Car and Driver on Volvo 1955-1986

BROOKLANDS PRACTICAL CLASSICS SERIES
PC on Austin A40 Restoration
PC on Land Rover Restoration
PC on Metalworking in Restoration
PC on Midget/Sprite Restoration
PC on Mini Cooper Restoration
PC on MGB Restoration
PC on Morris Minor Restoration
PC on Sunbeam Rapier Restoration
PC on Triumph Herald/Vitesse
PC on Triumph Spitfire Restoration
PC on VW Beetle Restoration
PC on 1930s Car Restoration

BROOKLANDS MOTOR & THOROGHBRED & CLASSIC CAR SERIES
Motor & T & CC on Ferrari 1966-1976
Motor & T & CC on Ferrari 1976-1984
Motor & T & CC on Lotus 1979-1983

BROOKLANDS MILITARY VEHICLES SERIES
Allied Mil. Vehicles No.1 1942-1945
Allied Mil. Vehicles No.2 1941-1946
Dodge Mil. Vehicles Col. 1 1940-1945
Military Jeeps 1941-1945
Off Road Jeeps 1944-1971
Hail to the Jeep
US Military Vehicles 1941-1945
US Army Military Vehicles WW2-TM9-2800

BROOKLANDS HOT ROD RESTORATION SERIES
Auto Restoration Tips & Techniques
Basic Bodywork Tips & Techniques
Basic Painting Tips & Techniques
Camaro Restoration Tips & Techniques
Custom Painting Tips & Techniques
Engine Swapping Tips & Techniques
How to Build a Street Rod
Mustang Restoration Tips & Techniques
Performance Tuning - Chevrolets of the '60s
Performance Tuning - Ford of the '60s
Performance Tuning - Mopars of the '60s
Performance Tuning - Pontiacs of the '60s

CONTENTS

Page	Title	Source	Date
5	1963 Corvette Sting Ray Sport Coupe Road Test	Car Life	Dec. 1962
10	Corvette Sting Ray	Car	Dec. 1962
12	Corvette Sting Ray Road Research Report	Car and Driver	April 1963
18	Corvette Sting Ray Road Test	Motor Trend	May 1963
24	The Lightweight Racing Sting Ray Track Test	Auto Sports	July 1963
26	Testing the Sting Ray Road Test	Hop-Up	Sept. 1963
32	Chevrolet Corvette Sting Ray	Motor	Dec. 11 1963
36	Sting Ray Automatic Road Test	Road & Track	March 1964
40	Corvette GS vs. Scarab-Chevy Comparison Test	Sports Car Graphic	March 1964
44	Chevrolet Corvette Sting Ray Road Test	Autosport	March 13 1964
46	Sting Ray Road Test	Motor Car Illustrated	May 1964
52	Corvette vs. Cobra Comparison Test	Car Life	Aug. 1964
55	Corvette Engineering for 1965	Motor Trend	Oct. 1964
58	Corvette Sting Ray Road Test	Car and Driver	March 1965
61	Corvette Sting Ray Mk.IV	Sports Car Graphic	March 1965
64	Corvette Sting Ray Road Test	Motor Trend	April 1965
70	Jaguar XKE vs. Corvette Sting Ray Comparison Test Road Test		May 1965
78	Corvette Sting Ray 427 Road Test	Car and Driver	Nov. 1965
82	427 Sting Ray Road Test	Sports Car Graphic	Dec. 1965
86	Corvette 427 Road Test	Motor Trend	March 1966
89	Corvettes Road Test	Car Life	Aug. 1966
97	Quickest, Fastest, Stoppingest 'Vette Yet! Road Test	Cars	June 1966
102	1967 Corvette Road Test	Road & Track	Feb. 1967
106	Corvette Grand Sports!	Car and Driver	April 1967
111	What Price Glory? Road Test	Cars	May 1967
117	Corvette Sting Ray, 36,000 Miles Later	Road & Track	Feb. 1967
120	Shelby GT 500 & 427 Sting Ray Road Test	Motor Trend	April 1967
126	Hottest 'Vette Yet! Road Test	Hot Rod	May 1967
130	The Sting Ray Emerges	Corvette – An American Classic	1978
143	The Second Generation	Corvette – An American Classic	1978
147	The Between Years: '64 to '67	Corvette – An American Classic	1978
160	Split Window Coupe	Road & Track Specals	1986
166	Chevrolet Corvette Sting Ray Profile	Classic & Sportscar	Sept. 1989

ACKNOWLEDGEMENTS

Brooklands Books publish source books. There is nothing original within their covers and their purpose is to make available to todays owners the road tests and other technical stories that were printed about a model when it was in production.

Amongst our 350 titles is one other on Corvettes — High Performance Corvettes 1983-1989 — which covers the more powerful examples produced during the eighties. We are currently working on the early Corvettes and later this year will publish a further Gold Portfolio dealing with the pre '63 cars.

Our books are printed in small numbers as works of reference for those that indulge in the hobby of automobile collecting and restoration. We exist firstly because there is a need by owners for this information and secondly because the publishers of the worlds leading automotive journals generously assist us by allowing us to include their copyright articles in these anthologies. We are indebted in this instance to the management of Autosport, Auto Sports, Car, Car and Driver, Car Life, Cars, Classic and Sportscar, Hop-Up, Hot Rod, Motor, Motor Car Illustrated, Motor Trend, Petersens Publications, Road & Track, Road & Track Specials, Road Test and Sports Car Graphic for their ongoing support.

R.M. Clarke

1963 CORVETTE Sting Ray Sport Coupe

Power steering, power brakes and Powerglide give this sports car even more popular appeal

Chevrolet's entrant in the runabout sweepstakes is no yearling—it's been around a good 10 years and helped nurture the whole movement. A true sports car (even purists now admit it), the Corvette enjoys a reputation unmatched by any other U.S.-built automobile and surpassed by only a few foreign-built ones. It earned this reputation the hard way: It won sports car races and, consequently, the sports car buyers.

But the same qualities which endeared it to the young and enthusiastic have also limited its sale to those not quite so young but considerably more affluent. Few members of the country club set seemed to care for the firm suspension, stiff steering and muscular performance; they bought 4-passenger Thunderbirds, softly sprung, power-operated and with more space to carry the golf clubs. While Corvettes sold at the satisfactory rate of 12,000-13,000 per year, the Thunderbird jumped to more than 80,000 per year.

Considering these facts, along with certain just criticisms by the sports car buff, Chevrolet engineers, when they laid the 1963 Corvette on paper, incorporated changes which would make the new models more acceptable at both ends of this peculiar market spectrum. Although they found the Thunderbird format incompatible with their perspective for the Corvette, they nonetheless carefully noted the reasons the 'Bird had reached its lofty sales perch. Consequently, the Corvette now is softer riding, easier steering, more striking in appearance and as comfortable as old bedroom slippers.

Although one might assume that the suspension needs for a competitive sports car would not be acceptable in a prestigious, personal runabout, the reverse is closer to the truth. Current road-racing practice is to use soft to very soft springing with very strong shock absorbing action: current passenger sedan design also specifies soft suspension but without quite so much shock absorber control. In fact, with ride rates (in lb./in. at the wheel) of 100 in. front and 116 in. rear, the new Corvette has slightly softer springing than does the Chevrolet sedan. Coupled with the independent wheel action, this provides unbelievable suppleness and tractive ability. Even without those certain tires which shoot out little hands to clutch a handful of paving, the new Corvette "Grrriippps the Road."

The front components of this new suspension are basically the ball jointed (Chevrolet calls them spherical joints because Ford had them first) members from the Chevrolet sedans, stamped parallel A-arms with coil springs sandwiched between. But the *piece de resistance* is out back. Here lurks the first modern fully articulated independent rear suspension produced (for passenger cars) in the U.S. It has a chassis-mounted differential and a 3-link location system. The axle halfshafts, which have universal joints at each end but no slip joints, serve as the upper member, a parallel lower arm pivots at hub-carrier and differential while the third member is a trailing radius rod. The radius arm transfers all driving and braking loads to the frame and extends back to form the hub carrier. The transverse leaf spring mounts behind the differential and connects to the radius arms by rubber-bushed rods. Chevrolet engineers say this cross-spring was used because of space limitations and because it lessens unsprung weight. Telescopic shock absorbers operate vertically off the radius arms. This system of rear suspension was first used on the Cisitalia Grand Prix car of 1947-48 and currently enjoys tremendous vogue among sports/road racing car builders.

A completely new frame was devel- ▶

CORVETTE Sting Ray

oped for the Corvette; a perimeter type with five cross-members. The body mounts on 12 pads but also has a complete subframe of stamped steel rails and posts. This subframe completely encompasses the interior of the coupe, supporting both roof and cowl panels, but necessarily is abbreviated in the convertible. The frame weight is 260 lb., approximately the same as previous Corvette frames, but beam strength has been improved and torsional strength boosted by 50%.

Wheelbase was reduced, from 102 to 98 in., on the new frame and overall dimensions slightly trimmed by the new bodies; it is now 1.4 in. shorter, 0.8 in. narrower and 2.4 in. lower than the 1962 model.

The new chassis and body combination has made a drastic improvement in the Corvette's weight distribution. Where before it was 53% on the front and 47% on the rear wheels, the situation is now reversed to 48% front and 52% rear. This is virtually perfect for competitive work and ideal, too, for just enjoyable driving. The overall weight of the new car is approximately the same as the old, with a total of 3048 lb. at the curb vs. 3080 in '62.

The power teams are unchanged from those offered in '62, the 260-bhp, 327-cu. in. V-8 with 3-speed transmission remaining as standard equipment. Optional V-8s of 300, 340 and 360 bhp are optional, as are the 4-speed all-synchromesh and Powerglide automatic transmissions. Just about the only change in this department was enlargement of the fuel injection system's plenum chamber to give quicker throttle response. Only the 250- and 300-bhp units may be ordered with power steering and Powerglide but power

INTERIOR OF COUPE is more orderly than before, features big, easy-to-read instruments. Luggage area behind seats is adequately spacious

brakes are optional with all engines.

The steering mechanism of the Corvette now has a recirculating ball system with a 16:1 ratio, a hydraulic damping device, provision for "quick" steering and an adjustable (in-and-out) wheel. The steering shaft is connected to the gearbox through a flexible, universal joint coupling which is splined to let the shaft move in or out. A clamp on the coupling locks the shaft in place and the range of adjustment is 3 in., which is enough to give most drivers a comfortable arm position.

Quicker steering is provided by an extra set of holes in each steering arm, closer to spindles; shifting the tie rod ball studs from the back holes to the front holes decreases the overall steering ratio from 19.6:1 to 17:1 and this change reduces steering wheel turns from 3.4 to 2.9 lock to lock.

The brakes, too, have come in for improvement. A self-adjusting feature has been added and the brakes enlarged to 328 sq. in. swept area from 1962's standard of 259. Sintered iron linings are optional but recommended (by Car Life) for the enthusiastic driver; for the would-be racer there's a "Special Performance" option which boosts the size to 334.3 sq. in., has metallic linings, a dual master cylinder with vacuum booster and larger diameter finned aluminum drums. These latter brakes are self-adjusting, too, but when the car stops going forward. The standard Corvette self-adjustment takes place on reverse stops.

Further options for competition are offered for the fastback coupe. These are heavy duty springs and shock absorbers, a stronger front stabilizer bar, knock-off cast aluminum wheels and a 36.5-gal. fuel tank. These probably will be available on the convertible at a later date.

The razzle-dazzle performance of the fuel-injected, 4-speed Corvette far overshadows that of the model Car Life has selected for its road test; however, we feel that the version we've tested here may become by far the most popular with the general public. It has a big, strong engine and an automatic transmission, a full complement of power assistants and a comfortable, controlled ride. The aerodynamic coupe body provides just the right touch to complete the ensemble; it is stylish and attractive yet efficient and well put together.

Driving such a Corvette is sheer delight. Despite its deceptive smoothness, it's a lusty performer only 2.8 sec. slower to 80 mph than the fuel-injected 4-speed convertible tested by *Road & Track* (October). Standing quarter-mile acceleration times are only a half-second slower, although the f.i. version has 60 bhp more and better final gearing (3.70:1 vs. 3.36:1). A brief comparison looks like this.

NOW-YOU-SEE-'em, now-you-don't lights retract electrically.

ENTRY IS easy into this coupe.

	300 bhp/auto	360 bhp/4-speed
0-40	4.2	3.4 sec.
0-60	7.2	5.9 sec.
0-80	13.0	10.2 sec.
0-100	22.9	16.5 sec.
¼-mile	15.5	14.9
Speed at end	86	95 mph

Neither of these cars were specially tuned. Both were pilot line models with a goodly number of test miles on

CORVETTE

the odometer. Consequently, both could reasonably be expected to do better with some careful attention. Another factor affecting straightline acceleration is this: the improved traction afforded by the new rear suspension plus rearward weight bias makes it extremely difficult to come off the line with wheels spinning. In fact, to get a little wheelspin on the f.i. car, the driver had to turn the engine up to 5000 rpm before dropping the clutch. Top speed is something else; the coupe is so good aerodynamically that it easily pulls the power peak and more. At 5500 rpm, the speed is 130 mph! The only limit here is pump-up of the hydraulic valve lifters and the quality of the tires. (We understand that a fuel-injected racing coupe has unofficially done better than 160 mph on the GM Proving Grounds test track—Ed.).

Tricky, twisting roads are this Corvette's meat. With its new suspension it seems to lock onto them, going precisely where directed and sticking to the tightest corners without the shadow of a doubt. Where the old Corvette had an annoying penchant for swapping ends when cornered vigorously, the new one just sticks and storms. This suspension is the best thing since gumdrops!

INDEPENDENT REAR suspension is a 3-link system; half-shaft is the upper arm, a parallel rod the lower and a radius arm the third member.

STEEL CAGE forms subframe of coupe body, provides pillars for door hinges and latches.

Indeed, we found very little to criticize in either version of the Corvette. Our two main objections centered on design features of the coupe: a) the bar down the center of the rear window makes it all but impossible to see out via the rearview mirror; and, b) getting luggage in and out of the adequate-sized compartment is difficult because it all has to go through the doors and over the seats—there's no trunk opening.

And, although we admit they do contribute to the overall streamlining of the car, we didn't really like the disappearing headlights. They seem a little too fussy for such an elegantly functional car. (Turning them on is accomplished by two switches, one to actuate the rotating motors and one to complete the headlight circuit.) No doubt, though, a lot of people will disagree with us about them. One thing they'll *all* agree on, ... this is the best Corvette yet!

CAR LIFE ROAD TEST

1963 CORVETTE
Sting Ray Sport Coupe

SPECIFICATIONS
- List price.................$4238
- Price, as tested............4720
- Curb weight, lb.............3048
- Test weight.................3398
- distribution, %............48/52
- Tire size...................6.70-15
- Tire capacity, lb @ 26 psi..4460
- Brake swept area............328
- Engine type.................V-8, ohv
- Bore & stroke...............4.0 x 3.25
- Displacement, cu in.........326.7
- Compression ratio...........10.5
- Carburetion.................1 x 4
- Bhp @ rpm...................300 @ 5000
- equivalent mph.............110
- Torque, lb-ft...............360 @ 3200
- equivalent mph.............75

EXTRA-COST OPTIONS
Power steering, power brakes, power windows, Powerglide transmission, 300-bhp engine, wsw tires, radio.

DIMENSIONS
- Wheelbase, in...............98.0
- Tread, f and r..............56.3/57.0
- Over-all length, in.........175.3
- width.......................69.6
- height......................49.8
- equivalent vol, cu ft.......351
- Frontal area, sq ft.........19.3
- Ground clearance, in........5.0
- Steering ratio, o/a.........17.0
- turns, lock to lock.........2.9
- turning circle, ft..........n.a.
- Hip room, front.............2 x 20.5
- Hip room, rear..............n.a.
- Pedal to seat back..........40.8
- Floor to ground.............7.5
- Luggage vol, cu ft..........n.a.
- Fuel tank capacity, gal.....20.0

GEAR RATIOS
- 4th (   ), overall
- 3rd (1.00).................3.36
- 2nd (1.76).................5.91
- 1st (1.76 x 2.10)..........12.4

PERFORMANCE
- Top speed (5500), mph......130
- Shifts, rpm @ mph (forced)
- 3rd (   )
- 2nd (   )
- 1st (5800)................78

ACCELERATION
- 0-30 mph, sec..............3.0
- 0-40.......................4.2
- 0-50.......................5.6
- 0-60.......................7.2
- 0-70.......................9.8
- 0-80.......................13.0
- 0-100......................22.9
- Standing ¼ mile............15.5
- speed at end...............86

FUEL CONSUMPTION
- Normal range, mpg..........12-16

SPEEDOMETER ERROR
- 30 mph, actual.............31.0
- 60 mph.....................58.8
- 90 mph.....................86.4

CALCULATED DATA
- Lb/hp (test wt)............11.3
- Cu ft/ton mile.............142.5
- Mph/1000 rpm...............23.5
- Engine revs/mile...........2555
- Piston travel, ft/mile.....1385
- Car Life wear index........35.4

PULLING POWER
- 70 mph, lb/ton.............450
- 50.........................off scale
- 30.........................off scale
- Total drag at 60 mph, lb...115

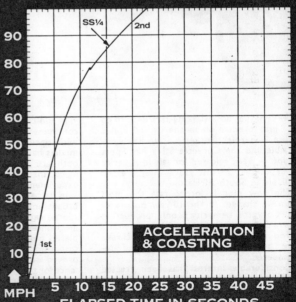

ACCELERATION & COASTING

NEW MODELS

CORVETTE STING RAY

Two striking cars — the Chev. Corvette Sting Ray convertible (foreground) and sport coupé.

★

The new Sting Ray chassis features a stronger frame, independent suspension all round and the differential fixed to the frame.

TWO of the most striking cars ever to go into production in the United States — the Chevrolet Corvette Sting Ray Convertible and the Corvette Sting Ray Sport Coupé — were released for sale in America on September 28 this year. They are not available in South Africa.

The new models incorporate the most dramatic changes in the 10-year history of the Corvette, America's only production sports car. For the first time since the car's introduction in 1953, both body and chassis have been completely restyled and re-engineered.

Considerable interest has been created by the addition of a second model Corvette — an exciting "fast-back" sports coupé for customers desiring closed car comfort with sports car appeal. "Fast-back" is the American name for the Gran Turismo style of car.

In addition to the sports coupé, Corvette continues to offer the convertible model with optional hardtop attachment. Both models have identical front end styling, and both are known as the "Sting Ray" — a name introduced on a recent Corvette experimental car.

In its tenth anniversary year, Corvette is broadening its customer appeal. To the sports car enthusiast it offers a host of mechanical improvements, including a completely new chassis, four-wheel independent suspension, self-adjusting brakes and improved steering, and for driving convenience for the first time optional power steering, power brakes and air-conditioning.

An intriguing feature on both models is retractable headlamps. Invisible until the driver presses a switch, the power-operated twin headlamp units swing up from the leading edge of the airfoil-shaped front end. A second switch turns them on.

The 1963 Corvettes are two inches lower and four inches shorter in wheelbase than the 1962 models. Overall height is 49·8 inches, wheelbase 98 inches, overall length 175·3 inches and the weight of the sports coupé is 3,015 lb., while the weight of the convertible is 3,037 lb.

The new Corvettes carry more weight on the rear. For the first time in contemporary American design, a front engine car carries more weight on the rear wheels than on the front, giving better rear wheel traction. The self-adjusting brakes have 18 per cent more bonded lining area than the previous models.

The steering wheel can be adjusted at the dealership to suit the driver's personal preference. There is a new ball-race steering gear and relay-type linkage with built-in hydraulic damper, which isolates road shock.

A wide variety of Corvette engines will continue to be offered with horse-power ranging from 250 to 360. There is a new fuel injection system on the 360 b.h.p. Ramjet Fuel Injection V8. Delcotron generators are standard equipment. (The Delcotron is a new compact and lightweight high-efficiency A.C. generator introduced recently by the Delco-Remy Division of General Motors.)

Three-speed, four-speed and Powerglide automatic transmissions with floor-mounted controls are available on the new Corvettes, plus a wide range of power train options, including positraction axles.

The fibreglass bodies of the 1963 models have additional steel reinforcing members. Despite their aerodynamic form, they offer greater interior room and better visibility.

THE WINNAH

The new Corvette Sting Ray has won the coveted CAR LIFE AWARD FOR ENGINEERING EXCELLENCE

Forgive our lack of modesty here, but we agree one hundred per cent with the editors of CAR LIFE. They think the new three-link independent rear suspension gives the car handling that's far and away the best thing ever to come from Detroit. So do we. They think the performance is on a par with any production sports car ever built. So do we and, we might add, so will you. Unfortunately, not everyone has had a chance to drive one of the new ones yet, because demand has exceeded production, but when your chance comes, you won't believe it! You've never driven a sports car that rides so well, yet handles so beautifully in the bargain. You've never sat in a car that'll turn so many heads and cause so much comment among the less fortunate drivers you pass. This car is a winner! And you'll share CAR LIFE's enthusiasm by the time you've hit forty miles per hour and second gear!... Chevrolet Division of General Motors, Detroit 2, Michigan.

NEW CORVETTE STING RAY BY CHEVROLET

Waiting lists of great length and duration for the Corvette Sting Ray at all Chevrolet dealers' are the best proof of the public's acceptance of the new model. We hailed the car's technical advances with great enthusiasm (*Oct. C/D*) after our brief test drives last fall.

Now it's time for an exhaustive report on America's leading grand touring car (which many drivers think of only as a sports car). We chose the 300-bhp version of the coupé, because it seems to enjoy some market preference over models equipped with the 250-, 340-, or fuel-injection 360-bhp engines.

However, the key to the personality of the Corvette Sting Ray lies neither in the power available nor in the revised styling, but in the chassis. Up to now the Corvette has been struggling to rise above a large number of stock components, notably in the suspension, where their presence created all kinds of problems that re-

quired extensive modifications for any competition use beyond normal road rallies. The new all-independent suspension has completely transformed the Corvette in terms of traction and cornering power, but it still has some faults. The standard setup on the test car seemed a bit more suitable for race tracks than for fast back-road motoring. A rigid front anti-roll bar in combination with a relatively stiff transverse leaf spring in the rear reduces the resilience and independence of the suspension of each wheel with the result that even on mildly rough surfaces the car does not feel perfectly stable. On bumpy turns it's at its worst, veering freely from one course to another, making high-frequency corrections s.o.p., but on a smooth surface it comes incredibly close to perfection. Cornering stability under conditions permitting minimal wheel deflections is remarkable, and an initial feeling of pleasant surprise

At long last America has a formidable weapon to challenge

CAR and DRIVER ROAD RESEARCH REPORT:
Corvette Sting Ray

rises to sheer astonishment when one discovers that the car can be taken off the predetermined line with ease and still complete the turn in perfect balance.

There is some understeer but the car has such a tremendous power surplus, even with the next-to-bottom engine option, that the tail can be slung out almost any old time, and after a while throttle steering seems the natural way of aiding the car around a curve. This is so easy to do that a newcomer to the car can master it in half an hour of fast driving.

Given surface roughness, the rear end becomes skittish. We experienced this with a full tank as well as one almost empty, indicating that normal loads don't appreciably affect its behavior in this respect.

One of our test cars had the new Saginaw power steering, three turns lock to lock with enough road feel to satisfy the most critical tester and observer, while eliminating all difficulties of parking and maneuvering in tight spaces. We also tested a car with manual steering, and found it so light in comparison with previous Corvettes that there can be no conceivable need for power assistance. While the power system is every bit as good as those used by Rover and Mercedes-Benz in terms of feedback and road feel, it seems strange that Chevrolet should get around to introducing it when there is no longer any need for it. The three-spoke wheel is steeply raked (15° 23′) as on previous Corvettes, and its relatively thin rim offers a good grip. The entire semi-circle between nine and three o'clock is free of spoke attachments, providing a clean hold for any but the most eccentric drivers. The steering column has a three-inch adjustment for length but our test drivers all kept the wheel in its foremost (bottom) position while making the most of seat-adjustment possibilities. There

Europe's fastest grand touring cars on their home ground

CORVETTE STING RAY

are four inches of fore-and-aft travel but backrest angle is variable only by setting screws at its floor abutments. In addition, there are three seat-height positions with a total span of 1.24 inches.

The result is a range of adjustment adequate to let our test drivers (ranging in height from five-seven to six-four) find a nearly ideal seating position. Maximum effective leg room (to the accelerator) is 43.7 inches and the maximum vertical height from the seat to the headlining is 33 inches. In view of the over-all height of only 49.8 inches, this is a good example of the care that has gone into designing the living quarters of the new Corvette Sting Ray.

As the engine and drive train are offset one inch to the right to provide wider leg room for the driver, he sits facing exactly in the direction he is going, with the pedals straight in front of him. The accelerator is nicely angled for normally disposed feet, but the clutch pedal has a rather excessive travel. With standard adjustment, you cannot release it without taking your heel off the floor, causing a bit of annoyance in traffic.

Instead of a fly-off handbrake, the Corvette has a T-handle under the instrument panel labeled "Parking Brake"—one of the few features of the new model which reminds you of its relationship with Chevrolet's mass-produced sedans.

Compared with previous Corvettes, the Sting Ray is improved in almost every imaginable respect: performance, handling, ride comfort, habitability and trunk space. The trunk is only accessible from inside the car, however, since the tail is full of fuel tank and spare wheel, but the storage space behind the seats is even larger than outside dimensions indicate. A third person, sitting sideways, may come along for short rides, but will soon feel cramped from lack of headroom. An occasional extra passenger will actually be better off sitting on the console between the seats and sharing legroom with the shotgun rider.

Having driven the Corvette Sport Coupé in all kinds of weather conditions, we found the heater and defroster units eminently satisfactory. The heater fan has three speeds, and air entry is variable by a push-pull control. Warm-up is not extremely rapid but seems to be faster than average. The body proved absolutely draft-proof and water-tight.

We liked the ball-shaped interior door handles but were not convinced of the advantages of the wheel-type door lock buttons. A minor complaint is the location of the window winders, as you cannot set your knee against the door panel for bracing on a sharp turn without coming in contact with the window handle.

Brakes have long been a sore point with Corvettes, and further advance has now been made without taking the full step of going to disc brakes (which the car really deserves). The Delco-Moraine power brakes have 11-inch steel drums cast into the wheel rims, with 58.8% of the braking force being directed to the front wheels. Sintered iron brake linings are optional and will certainly be found necessary for anyone planning to race, as fade is easily provoked with the standard linings, although the cooling-off period required to restore full efficiency is very short.

Chevrolet is prepared for a fair-sized demand for special performance parts, but has restricted their application to the structurally stronger Sport Coupé. The sintered-iron heavy-duty brake system also includes vented backing plates and air scoops and a dual-circuit master cylinder. There is a heavy-duty anti-roll bar, heavy-duty front and rear shock absorbers, aluminum wheels with knock-off hubs, and a 36-gallon fuel tank. The brake mechanism, in contrast to that fitted as standard, automatically adjusts the brakes when applied during *forward* motion. To be ordered, this special performance kit (RPO ZO6) also requires the 360-bhp engine, the four-speed Warner T-10 gearbox and a Positraction limited-slip differential.

Race preparation of the 327-cubic-inch Corvette engine has been thoroughly treated by Bill Thomas in an article for the *Corvette News* (Volume 5 No. 3), a GM publication invaluable to both the active Corvette competitor and his "civilian" counterpart. For information, readers are advised to write to *Corvette News*, 205 GM Building, Detroit 2, Michigan.

For all kinds of non-competitive driving, the 300-

Luggage space is surprisingly roomy but central window partition ruins rear view.

A ventilated fuel filler cap is reached through lid. Gear positions are labeled.

Body was wind-tunnel tested but many

bhp version gives more than ample performance for anyone, with our average standing-quarter-mile time at 14.4 seconds. This was achieved with the "street" gearbox and an axle ratio which limits top speed to about 118 mph, a combination which results in extreme top-gear flexibility as well. Top-gear starts from standstill to limit wheelspin present no problem with regard to stalling, but detonations were inevitable.

Fiberglass bodies usually have peculiar noises all their own but the Corvette was remarkably quiet, no doubt due to the steel reinforcement surrounding the entire passenger compartment. The car is also notable for low wind noise and high directional stability. Engine noise is largely dependent on the throttle opening —it will respond with a roar to a wiggle of the toe if you're wearing light shoes, and this holds true within an extremely broad speed range. Top-gear acceleration from 50 to 80 is impressive indeed, both in sound and abdominal effects.

In this connection, the gear lever has a set of speeds at which it vibrates and generates a high-pitched rattle (this is in the lever itself and not in the reverse catch), and there are intermittent peculiar noises from the clock, probably when it rewinds itself.

The now-familiar Warner T-10 gearbox has faultless synchromesh and when fully broken in can be as light as cutting butter. One interesting aspect of its operation is the fact that the owner's handbook specifies double-clutching for down-shifts.

We are in complete agreement with this recommendation, over which there has been some controversy. Some people feel that double-clutching will wear out the synchromesh. This can be true only if on downshifts the engine is accelerated so much that the synchromesh has to work harder than it would with a single-clutch change, a situation which does not seem to occur very often.

While we agree that the Buick Riviera, for example, is the kind of car where automatic transmission has a function, we cannot see its place in the Corvette and our testing was done exclusively on a pair of manual-shift cars, one with power steering and one without, neither with Positraction limited-slip differential, which perhaps should be standard equipment on this car.

As the majority of new Corvettes are built with four-speed transmissions, it is hard to understand why the three-speed remains listed as standard equipment. We can see no reason for even continuing to offer it, and recommend that both the Powerglide and the three-speed manual gearbox be dropped. This would let Chevrolet standardize the wide-ratio four-speed transmission throughout and make the close-ratio version optional for the 340- and 360-bhp models.

Our testers preferred the car with the fewest automatic "aids," and probably most of our readers will, too. That keen drivers prefer manual controls is not baffling at all—except possibly to advanced research personnel who forget that nowhere else can they get an effective 180-pound corrective computer which can be produced at low cost by unskilled labor.

Vastly more practical than any previous Corvette, the Sting Ray Sport Coupé appeals to a new segment of buyers who would not be interested in a convertible, and production schedules at the Saint Louis assembly plant have been doubled from the 1962 model's. As an American car it is unique, and it stands out from its European counterparts as having in no way copied them but arrived at the same goal along a different route. Zora Arkus-Duntov summed it up this way: "For the first time I now have a Corvette I can be proud to drive in Europe." We understand his feelings and are happy to agree that the Sting Ray is a fine showpiece for the American auto industry, especially since it is produced at a substantially lower price than any foreign sports or GT car of comparable performance. **C/D**

Directional and parking lights are part of bumper design but the retractable headlamps are concealed for daytime driving.

object to superfluous decoration by emblems and dummy vents.

The Corvette is perhaps best looking from behind, and this is a view that drivers of other cars will soon become used to.

Road Research Report
Chevrolet Corvette Sting Ray Sport Coupe

Manufacturer: Chevrolet Motor Division
General Motors Corporation
Detroit 2, Michigan

Number of U.S. dealers: 7,000 (approximately)
Planned annual production: 16,000

PRICES
Basic price $4,252

OPERATING SCHEDULE
Fuel recommended ....Premium (99-101 Octane)
Mileage ....10-18 mpg
Range on 20-gallon tank ....200-360 miles
Oil recommended   Single grade   Multi-grade
32° F and over.............SAE 20 or 20W   SAE 10W-30
0° F ....SAE 10W   SAE 10W-30
below 0° F ....SAE 5W   SAE 5W-20
Crankcase capacity ....5 quarts
Change at intervals of ....6,000 miles
Number of grease fittings ....10 (9 with manual steering)
Most frequent maintenance.......Lubrication at every 6,000 miles

ENGINE:
Displacement ....327 cu in, 5,370 cc
Dimensions.......................8 cyl, 4.00-in bore, 3.25-in stroke
Valve gear: ......Pushrod-operated overhead valves (hydraulic lifters)
Compression ratio ....10.5 to one
Power (SAE) ....300 bhp @ 5,000 rpm
Torque ....360 lb-ft @ 3,200 rpm
Usable range of engine speeds ....600-5,500 rpm
Carburetion ....Single four-throat Carter WCFB carburetor

CHASSIS:
Wheelbase ....98 in
Track ....F 56.3 in, R 57.0 in
Length ....175.3 in
Ground clearance ....7.5 in
Suspension: F: Ind., coil springs and wishbones, anti-roll bar
R: Ind., lower wishbones and unsplined half-shafts acting as
locating members, radius arms and transverse leaf spring
Steering ....Saginaw recirculating ball with power assistance
Turns, lock to lock...3
Turning circle diameter between curbs........................36 ft
Tire size ....6.70 x 15
Pressures recommended ....F 24, R 24 psi
Brakes...Delco-Moraine 11-in drums front and rear, 328 sq in swept area
Curb weight (full tank)......................................3,180 lbs
Percentage on the driving wheels................................53

DRIVE TRAIN:
Clutch...........................Borg & Beck 10-in single dry plate

Gear	Synchro	Ratio	Step	Over-all	1,000 rpm
Rev	No	2.61		8.78	−9.0
1st	Yes	2.54	34%	8.52	9.3
2nd	Yes	1.89	25%	6.36	12.4
3rd	Yes	1.51	51%	5.08	15.6
4th	Yes	1.00	—	3.36	23.5

Final drive ratio...3.36 to one

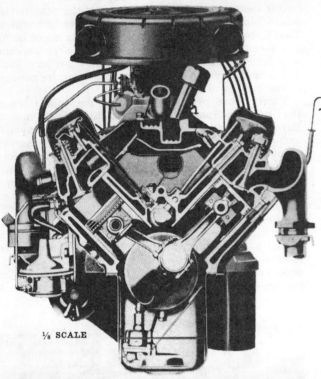

1/8 SCALE

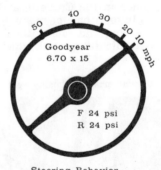

Steering Behavior
Wheel position to maintain 400-foot circle at speeds indicated.

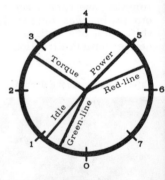

Engine Flexibility
RPM in thousands

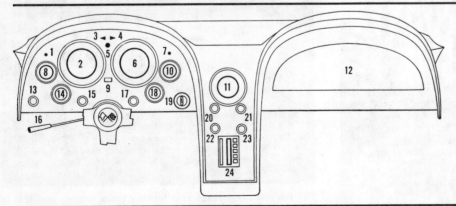

(1) Turn signal warning light (left); (2) Speedometer and odometer; (3) Warning light for headlights on in closed position; (4) Parking brake warning light; (5) High beam warning light; (6) Tachometer; (7) Turn signal warning light (right); (8) Water temperature gauge; (9) Trip odometer; (10) Oil pressure gauge; (11) Clock; (12) Glove box; (13) Light switch; (14) Ammeter; (15) Windshield wiper and washer; (16) Turn signal lever; (17) Cigarette lighter; (18) Fuel gauge; (19) Ignition key and starter; (20) Heater fan and fresh air control; (21) Defroster control; (22) Radio volume and tone control; (23) Radio tuning selector; (24) Radio dial.

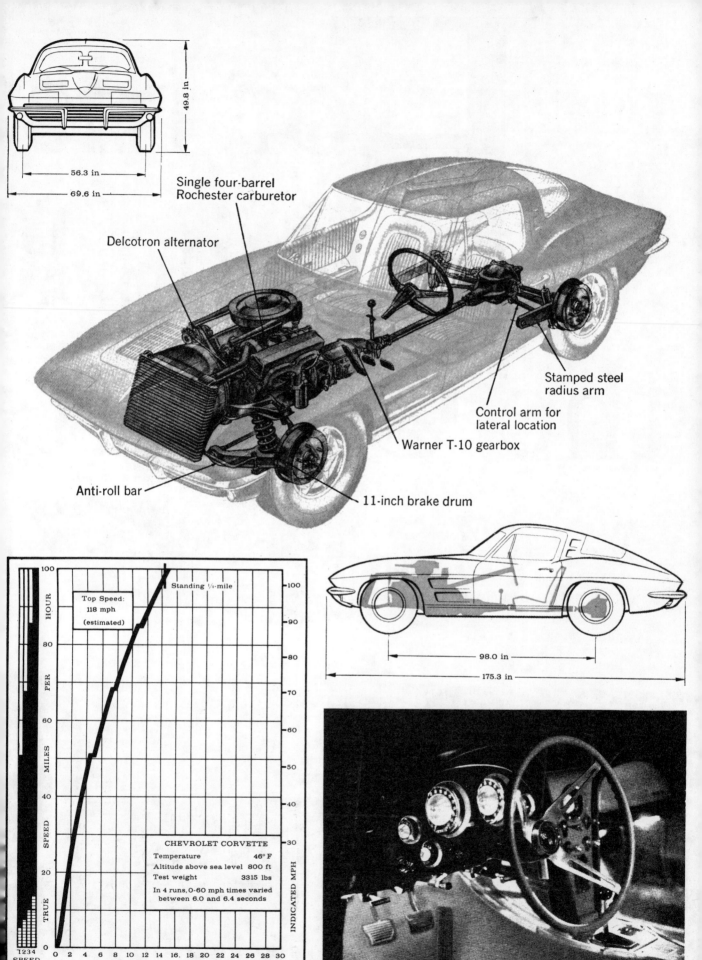

For the F in its 10 the Corvette

MT Road Test

by Jim Wright, *Technical Editor*

first time -year history, STING RAY is...

continued

... in such demand that the factory has had to put on a second shift and still can't begin to supply cars fast enough. The waiting period is at least 60 days, and dealers won't "deal" a bit on either coupes or roadsters. Both are going for the full sticker price, with absolutely no discount and very little (if any) over-allowance on trade-ins.

This is a healthy situation for Chevrolet, and we're happy to see the 'Vette get the public acceptance we've always felt it deserved. Yet, after giving the new one a thorough shakedown, we can't help but let our thoughts stray back to last year's road test and one statement in particular we made.

It had to do with how the factory has never really made any big profits on the Corvette, but that Chevy brass was more than satisfied as long as it carried its performance image and prestige over to the bread-and-butter lines. We also ventured an opinion that as long as the factory kept building the car on this basis it would be a great automobile, but if they ever put it on a straight dollar-profit basis, the Corvette would probably be ruined.

Well, they haven't ruined it yet, but our test car (as well as several others we've checked out) showed definite signs that the factory might be getting more interested in dollars than in prestige. From the important styling and all-around performance angles, the new Sting Ray is an even greater car than its predecessors. But for a car that sells in the $4500-6000 range, it doesn't reflect the degree of quality control we feel it should.

To begin with, there still seems to be some difficulty in manufacturing a really smooth fiberglass body. While this isn't too apparent in a light-colored car, it becomes all too noticeable in some of the darker ones. When the light hits these from almost any angle, there's a definite rippled effect. The doors on our test car had gaps around them that were wider than they should have been. The doors didn't line up too well, either. This was also true of the hood.

In the past, we've always been impressed with interior trim in most GM products. The moldings usually look as if they've been very carefully designed for a precision fit. To our eye, the Sting Ray coupe's interior had an unfinished look. Not that the upholstery and carpet materials weren't top grade — they were — but the various door and window moldings aren't too well designed in the first place, and it doesn't take much laxity of quality control on the assembly line to make them look really bad. While these aren't earth-shaking faults or defects, and have absolutely nothing to do with the operation of the car, they are of the sort that a discerning owner and driver will be constantly aware of.

For the first time, the Corvette is available with power steering and power brakes. We didn't have either on the test car, but we've driven several set up this way. Combined with one of the smaller engines and Powerglide transmission, these power accessories make the Sting Ray docile enough for little

(ABOVE) *Said to be the result of extensive wind-tunnel testing, Sting Ray's boat tail may give less wind resistance, but it definitely hampers vision.*

(LEFT) *Rochester fuel injection has been redesigned again this year. The 327-cubic-inch V-8 delivers 360 hp at 6000 rpm, has good torque output.*

PHOTOS BY PAT BROLLIER

old ladies or any other types interested in a nice, quiet, Sunday-go-to-meeting car.

The basic power trains are carried over from last year and include four engines of 250, 300, 340, and 360 hp. Powerglide's available with either the 250- or 300-hp engines. Basic transmission with all engines is a three-speed manual, with four-speed available optionally. The 340- and 360-hp engines use the close-ratio (2.20-to-1 low gear) four-speed. Six rear-axle ratios are available and include the 3.36 (standard with three-speeds and Powerglide), 3.70 (standard with four-speed transmission), plus other gearsets of 3.08, 3.55, 4.11, and 4.56 to 1.

The MOTOR TREND test car was equipped with the fuel-injection, 360-hp mill, four-speed, and 3.70-to-1 Positraction rear axle. Sintered-iron brake linings and heavy-duty suspension completed the option package.

On a straight acceleration basis, there's very little difference between last year's car and the new one. Our quarter-mile times are within fractions of what they were last year. The only real difference is that the new one doesn't have quite the wheelspin (with stock tires) that the old rigid-axle car had. The 0-30, 0-45, and 0-60-mph steps averaged 2.9, 4.2, and 5.8 seconds, while our average time through the quarter-mile traps was 102 mph, with a 14.5-second ET. Top speed was an honest 130 mph, with the tachometer reading 6000 rpm. A course longer than the Riverside Raceway backstretch would've produced something very close to the Sting Ray's theoretical top speed of 140-142 mph (with 3.70 gears), because the engine was still winding when we had to back off. The 360-hp engine is set up like any

(ABOVE RIGHT) *Optional brake linings of sintered iron don't require special drums, are practically impossible to fade.*

(RIGHT) *Spare location isn't so awkward as it looks, although it points up the importance of carrying overalls in the car.*

NEW INDEPENDENT REAR SUSPENSION AND 49/51 WEIGHT DISTRIBUTION REQUIRE SLIGHTLY DIFFERENT CORNERING TECHNIQUES THAN BEFORE.

WHEN VIEWED IN PROFILE, AERODYNAMICALLY STYLED STING RAY BODY SHOWS A CLEANLINESS OF LINE UNMATCHED BY ITS ANCESTORS.

 *continued*

well-designed racing engine, and it's very strong throughout the entire rpm range.

This is one of the few high-output engines that can deliver decent gas mileage without being babied. Out on the highway, we averaged slightly better than 18 mpg for one trip where we didn't go above the legal limits. On another trip where the speedometer stayed above 75 and 80 mph a good deal of the time, we saw 16.3 mpg. Whipping around town produced a 13.6-mpg average. For over 700 miles of all types of driving, the Sting Ray averaged 14.1 mpg.

A lot of this is due to the low weight the engine has to pull around, but the excellent Rochester fuel-injection unit also has something to do with it. Like any good injection unit, it can constantly correct itself to suit different humidity, temperature, and altitude conditions. All in all, it's a completely efficient system. It doesn't seem to be temperamental, either. Care does have to be taken to keep the fuel filters clean and operating because they're quite susceptible to dirt.

We'd recommend the metallic brake option to any buyer, regardless of the engine he's getting in his new Corvette. It's very reasonably priced ($37.70), and is unmatched for efficiency. The brakes in the test car were used very hard at the end of several high-speed runs and showed very little tendency to fade. It's true that they require more pedal pressure to operate and are a trifle noisy on cold mornings, but once they get warmed up they're excellent. If a woman is going to be driving the car a lot, power assist can be added to keep her from developing bulging muscles in the brake leg. Self-adjusting brakes are now standard equipment, and unlike systems in most other cars, they'll adjust as the car moves forward. The brakes in our test car pulled the Sting Ray down to quick straight-line stops time and again without any sudden locking of the wheels and without apparent fade. Several stretches of mountain roads showed that they could stand up to prolonged hard use without failure.

For all-out competition there's a special performance brake option that includes a little more effective brake lining area, bigger finned drums, power assist, and a divided output master cylinder (separate system for front and rear), and provisions for cooling. Combined with the optional cast-alloy wheels, this setup gives the competition Corvette braking power on par with many of its disc-braked competitors.

Since we had a complete analysis of the new suspension system in the January issue, we won't go into the theory of it here. In practice, it's far in advance, both in ride and handling, of anything now being built in the United States. It's completely comfortable without being mushy, and it takes a large chuck hole to induce any degree of harshness into the ride. Sudden dips, when taken at speed, don't produce any unpleasant oscillations, and the front and rear suspension is very hard to bottom. There's very little pitch noticeable in the ride, even though the 'Vette is built on a fairly short (98-inch) wheelbase. At high cruising speeds — and even at maximum speeds — nothing but an all-out competition car will equal it in stability. We drove it under some pretty windy conditions and didn't notice any adverse effects from crosswind loading.

We thought the old model cornered darn well, but there's no comparing it to this new one. It does take a little different technique, but once the driver gets onto it, it's beautiful. Since the 49/51 per cent front-to-rear weight distribution,

---

CORVETTE STING RAY
2-passenger coupe

OPTIONS ON CAR TESTED: 360-hp fuel-injection engine, 4-speed transmission, Positraction rear axle, sintered-iron brake lining, power windows, radio, whitewalls
BASIC PRICE: $4393.75
PRICE AS TESTED: $5321.70 (plus tax and license)
ODOMETER READING AT START OF TEST: 4012 miles
RECOMMENDED ENGINE RED LINE: 6500 rpm

PERFORMANCE

ACCELERATION (2 aboard)
0-30 mph.............................2.9 secs.
0-45 mph.............................4.2
0-60 mph.............................5.8
Standing start ¼-mile 14.5 secs. and 102 mph
Speeds in gears @ 6500 rpm
1st ....65 mph   3rd ....110 mph
2nd ....88 mph   4th......130 (actual) @ 6000 rpm
Speedometer Error on Test Car
Car's speedometer reading 30 45 51 61 72 82
Weston electric speedometer ...30 45 50 60 70 80
Observed miles per hour per 1000 rpm in top gear ....21 mph
Stopping Distances — from 30 mph, 30 ft.; from 60 mph, 134 ft.

SPECIFICATIONS FROM MANUFACTURER

Engine
Ohv V-8
Bore: 4.00 ins.
Stroke: 3.25 ins.
Displacement: 327 cu. ins.
Compression ratio: 11.25:1
Horsepower: 360 @ 6000 rpm
Torque: 352 lbs.-ft. @ 4000 rpm
Horsepower per cubic inch: 1.1
Ignition: 12-volt coil

Gearbox
4-speed, close-ratio manual; floor shift

Driveshaft
One piece — open tube

Differential
Hypoid — semi-floating (Positraction)
Standard ratio: 3.70:1

Suspension
Front: Independent, with coil springs, unequal-length upper and lower control arms, direct-acting tubular shocks, and anti-roll bar
Rear: Independent, with fixed differential, transverse, 9-leaf spring, lateral struts and universally-jointed axle shafts, radius arms, and direct-acting tubular shocks

Steering
Recirculating ball
Turns: 3.4 or 2.92 lock to lock

Wheels and Tires
5-lug, steel disc wheels
6.70 x 15 4-ply
nylon tubeless tires

Brakes
Hydraulic, duo-servo, self-adjusting, with sintered-iron linings, cast-iron drums
Front: 11-in. dia. x 2.75 ins. wide
Rear: 11-in. dia. x 2.0 ins. wide
Effective lining area: 134.9 sq. ins.

Body and Frame
Full-length, ladder-type frame, with 5 crossmembers and separate body
Wheelbase: 98.0 ins.
Track: front, 56.3 ins.; rear, 57.0 ins.
Overall length: 175.3 ins.
Curb weight: 3150 lbs.

plus the independent rear suspension, gives the Sting Ray an inherent amount of oversteer, the driver will find that on fast corners the car will be doing most of the work through the corner instead of him powering it through.

At most speeds the coupe and the roadster are quite noisy. In addition to high engine and wind noise levels, the coupe picks up and amplifies quite a bit of road noise through the differential, which is rigidly mounted (although in rubber) to the frame. The extremely stiff, ladder frame and well-designed body, with its built-in steel bracing (on the coupe), keep body shake to an absolute minimum.

The Sting Ray is roomier than the older models, and quite a bit of luggage can be carried in the space provided. Some people will be unhappy without a deck lid, but it's really not too inconvenient to get to the storage space through the passenger compartment. The steering wheel is now adjustable, although not readily so from the driver's seat. It has to be done in the engine compartment with a simple wrench, but gives up to three inches of fore and aft adjustment.

The bucket seats offer more of a contour fit to the back and are very comfortable once you get used to the low seating position. A full array of instruments is functionally arranged on the dash within easy view of the driver's eyes. But their design is such that at night, with the instrument lights on, they're hard to read. The brushed aluminum backing of each gauge tends to glare. The rear window on the coupe is designed more for looks than practicality, and any decent view to the rear will have to be through an exterior side-view mirror. /MT

(ABOVE) *Disappearing headlights are shown in three stages. Actuated by electric motors, they allow the Sting Ray a wind-cheating frontal design.*

(ABOVE) *Storage area is surprisingly adequate, though some would prefer a rear opening for easier access. In a pinch, adult can sit here on short hop.*

(LEFT) *All the necessary instruments are included in a well grouped layout. Shift lever is easy to reach, and the linkage has strong and positive action.*

AUTO SPORTS TESTS LIGHTWEIGHT RACING STINGRAY Corvette

by Bill Muldoon

This has got to come in under the heading of SCOOPS as last weekend we had the opportunity of track testing the much talked about but seldom seen lightweight Stingray from the now defunct competition department of General Motors. As you might expect being an 1870 lb. car with all the competition goodies it is a bomb .... it's loud, quick, hairy and responsive.

The car is one of two built by Zora Arkus-Duntov and his crew, and has been under development for quite some time. After investing about $40,000 in the experiment, G.M. for reasons of their own decided to get out of the racing business. The car pictured on this page is owned by the Performance Engineering Division of Doane Chevrolet, Dundee, Ill. It was delivered to us at Meadowdale International speedway by PE's head man Anthony Bosura.

The car was painstakingly designed and built. Every part was constructed with an eye to keeping the weight down and as such carrier a code number and weight listing. The overall dimensions are identical with the standard Corvette but the frame and most other parts is tubular and hand welded. The windshield is glass but side and rear windows are plastic.

The body is fiberglass of a new hand-made, hand-laid variety. We couldn't get Bosura to admit it but we have reason to believe that the engine block is aluminum. Specially designed fuel injection is used.

It was a good day for car testing at Meadowdale ... sunny and clear and not too many practicing cars and drivers. Bosura fired up the hot Chevy engine and it sounded like a couple of Sherman tanks at play. Two chrome tuned exhaust stacks protrude from each side of the car to carry gases to just aft of the drivers cockpit. Seated in the Lightweight, we surveyed the interior. Very businesslike. Grouped aircraft-type instruments, 8000 RPM tach, 2000 MPH speedometer, Lightweight seats and well-placed pedals for the heel and toe bit.

I took a slow lap to become acquainted with the machine. Steering was quick ... two turns from lock to lock. The Close ratio gearbox was

Photos by Sports Car Photo Service

smooth and positive and well positioned. The first sensation however was the feeling of lightness. I had a feeling that with all its speed potential, coupled with its lightweight, the car would be skittish. I was right. As I increased speed for the next few laps, I came to realize that the disc brakes characteristically needed warming up before they operated to full effect. But they were good brakes. A shoulder harness is recommended, for there's a bit of flopping around on the part of the driver ... at least on Meadowdale's twisting circuit. The car has understeer, which is a tribute to its special suspension. The rear view mirror needs repositioning, as it blocks the drivers view.

So, in a moment of false confidence I punched it down Meadowdale's long pit straight and **really** took off. I was busy, believe me, with **both** hands. I couldn't keep my eye on the speedometer, but cranked the bomb to about 6500 RPM before backing off for Turn One. Then, I started to get the hang of it and had a peachy fine time.

Meadowdale is a course that demands torque. You spend most of your time in third gear and there are two hills to be climbed. The Stingray is a real torquer and literally gobbled up the turns, most importantly the short straights between turns. The car is a barnburner and should be driven by someone familiar with Corvettes. Once a guy gets the hang of it he's going to be mighty hard to beat because this car has plenty in reserve.

As to its future. The car was homologated by the FIA, as GM planned to build about 400 units to sell for about $10,000 each. This approval was withdrawn when GM decided to back away from racing. So the cars that were designed to compete in A Production against standard Corvettes, Cobras and Ferraris are now slotted in C Modified. They haven't a chance. Doane therefore is dropping in a dyno-tested 450 HP Chevy engine, eliminating production hardware demanded of FIA production cars, and thus will drop the weight to 1700 lbs. With this weight reduction, hotter engine, the signing of a well known and successful Corvette driver, Doane expects the car to go 7 MPH faster than the Chapparal and be ready to do this at Elkhart Lake June 22-23. Lots of luck fellas. You've got a tremendous automobile, but it's too fast for *me*.

AUTO SPORTS tests the General Motors experimental lightweight StingRay at Meadowdale Race Track. The trim car has a lightweight look. The car is owned by Dick Doane, Dundee, Ill., Chevrolet dealer.

Missing from this Corvette is the center bar which is seen on most production models. And unlike the production models, this one has a trunk into which the spare tire fits nicely.

Other unusual features of this StingRay are the gas tank which is filled from the side rather than the rear; louvers in the lower left corner are so thin they spring back at a touch; and the air scoup for cooling brakes is on the side.

Bill Muldoon pulls the StingRay through the Meister Brau bridge at Meadowdale, and then poses for the camera.

The car has an air spoiler on the front which deflects the air to bring the front end of the car down during high speed driving. Headlamps are trim with Plexiglas for lightness.

Anthony Bosura lowers the windows of the StingRay by means of a strap, since there is no mechanism for rolling it. The only area of the car that is glass is the windshield; Plexiglas has been used to contribute to the car's lightness.

testing the

THERE comes a time in the life of every car tester when he hates to return the car. The new Sting Ray is the kind of car you borrow and hate to return.

Corvette's current offering is virtually all new for '63 and has a host of new features that should appeal to performance-minded enthusiasts.

The chassis has been completely redesigned, and greater use is made of passenger car front-end components. The wheelbase has shrunk from 102 to 98 inches. The box section frame is quite stiff; Chevy claims something like 90 percent more rigidity for the coupe over last year's car. The weight distribution is now 45/51, which pays off in improved handling and ride.

The big news, aside from style, is in the rear end. Independent rear suspension, with the differential mounted on the frame, accounts for the big plus factor in performance and traction.

The rear whels are driven by a series of U-joints and stubby driveshafts. The differential is mounted on thick rubber pads, thus isolating noise and vibration. A variety of optional gear sets are available from 3.08 to the high 5's. Positraction is also available, and our test cars were so equipped.

The tried and true four-speed box is available with two sets of ratios. The bigger engines have the box with the 2.20 first gear. For driving pleasure and control in a car like the Sting Ray, we can't see anything less than a four-speed box. Of course, some people will still buy the car with a three-speed box or Powerglide, and they will never know what they're missing.

Not that the new Corvette doesn't have its faults. For one thing visibility from the rearview mirror isn't too good. Chances are you'd probably *hear* a motorcycle

STING RAY

BY ED PHILLIPS AND DON MARTIN

ABOVE: In this maneuver, the Ray makes a big splash in more ways than one as it neatly negotiates its way through a wet cinder underbase.

LEFT: Stoppability gets an AOK. Brakes were very effective and showed no sign of fade after repeated application.

ABOVE: Car had some shift on sharp turns at about 45 mph.

ABOVE: Independent rear bite made up for front lean.

ABOVE: Headlights, housed in swivel units, can be turned . . .
BELOW: . . . under when not in use, via switch beneath dash.
RIGHT: Center console conveniently locates most controls.

testing the STING RAY

ABOVE: Tackling winding road was child's play for the Ray.

Normal front-end attitude (as shown above) changes when gas pedal is punched, and the front end lifts (as shown below).

cop on your tail before you could see him.

Other shortcomings:

—The awkward position of the gas pedal is a source of undue discomfort.

—The interior finish did not come up to our expectations. It was comparable to that of a low-priced compact.

—The spare tire is difficult to get at.

—The coupe's luggage compartment, while offering more room than last year's, is right behind the seats. On quick turns and sudden stops, we were attacked from behind by the camera equipment we carried. Shock cord tie-downs should be supplied on a car of this type.

Up front is the fire-breathing 327-cube engine that has honked Chevy to a top spot in the performance drive.

We had two test cars, one with the 300-hp engine, power steering and power brakes. The other had the 340-hp mill with solids and a big AFB carb.

Naturally, the 340 was zippier and more potent, but the 300 was just the ticket for city driving. In both handling and hustle, it more than filled the bill.

Our quarter-mile times with the

The 340-hp Ray was fairly thirsty and averaged about 11.5 miles to gallon.

testing the STING RAY

340 averaged 13.9 e.t. and 109 mph —this was with cheater slicks and a 4.56 rear. No doubt a super-tuned model set up by a drag strip perfectionist would have shown better times. But the promise and potential were there.

A few small quirks aside, credit the Corvette Sting Ray with being one helluvanautomobile!

ABOVE: 327-cube mill spurs 340 horses as the Sting Ray smokes off the line. No wheel hop was encountered in this test.

RIGHT: Although handling was generally excellent, car tended to drift left as it tore loose from a standing start.

BELOW: Despite test-driver Don Martin's effort to keep car straight, it veered to the left, as rubber marks indicate.

"If there were people who doubted whether or not the large American companies knew how to build a real sports car or whether or not they were interested in building one, they all know differently now."

—Paul Frere, Auto, Motor und Sport

Noted European automotive journalist Paul Frere recently put a Corvette Sting Ray through its paces in a road test for the German magazine, Auto, Motor und Sport. The above is a summary of his findings. Below are a few detailed comments.

Speaking of Corvette's handling, Mr. Frere said, "The driving characteristics of the Sting Ray are by all means comparable with the best European sports cars. The suspension is soft but well adjusted, which gives you a high degree of driving comfort and a feeling of safety at high speeds on streets which are not particularly smooth.... The steering has to be called exemplary."

Mr. Frere found the Corvette transmission to his liking: "The shifting itself is an actual pleasure with the short, precise and smooth lever. All gears are absolutely foolproof synchronized and the shifting process takes place as fast as you can move the lever from one position to the other."

Of the Corvette Sting Ray's overall sports car capabilities, Mr. Frere concluded, "The Sting Ray in this respect is not second to any one of the best European sports cars.... It can compete with them in price, even in Europe, and it tops them all in one area which is very important to those drivers who do not use their sports cars for fun only: There is a wide-spread, well organized customer service available."

America hears a lot about European sports cars. It's interesting to hear what Europe thinks about our sports car.... Chevrolet Division of General Motors, Detroit, Michigan.

CORVETTE STING RAY

MAKE: Chevrolet TYPE: Corvette Sting Ray
MAKERS: General Motors Corporation of America.
CONCESSIONNAIRES: Lendrum and Hartman Ltd., 26 Albemarle Street, London W.1.

ROAD TEST ● No. 47/63

TEST DATA:

World copyright reserved; no unauthorized reproduction in whole or in part.

CONDITIONS: Weather: Dry, very little wind. (Temperature 60°–78°F., Barometer 29·60 in Hg.). Surface: Dry concrete and tarmacadam. Fuel: Super Premium grade pump petrol (101 Octane by Research Method).

MAXIMUM SPEEDS
Flying Quarter Mile
Mean of four opposite runs .. .. 146·6 m.p.h.
Best one-way time equals .. .. 147·0 m.p.h.
"Maximile" Speed: (Timed quarter mile after one mile accelerating from rest)
Mean of four opposite runs .. .. 141·6 m.p.h.
Best one-way time equals .. .. 142·8 m.p.h.
Speed in Gears (at 6,300 r.p.m.)
Max. speed in 3rd gear .. .. 106 m.p.h.
Max. speed in 2nd gear .. .. 84 m.p.h.
Max. speed in 1st gear .. .. 62 m.p.h.

ACCELERATION TIMES
From standstill
0–30 m.p.h. .. .. 3·0 sec.
0–40 m.p.h. .. .. 4·1 sec.
0–50 m.p.h. .. .. 5·1 sec.
0–60 m.p.h. .. .. 6·2 sec.
0–70 m.p.h. .. .. 8·2 sec.
0–80 m.p.h. .. .. 10·2 sec.
0–90 m.p.h. .. .. 13·3 sec.
0–100 m.p.h. .. .. 15·6 sec.
0–110 m.p.h. .. .. 19·6 sec.
0–120 m.p.h. .. .. 23·6 sec.
Standing quarter mile .. .. 14·6 sec.

On upper ratios

	Top gear	Third gear
10–30 m.p.h.	—	4·0 sec.
20–40 m.p.h.	5·2 sec.	3·6 sec.
30–50 m.p.h.	4·7 sec.	3·4 sec.
40–60 m.p.h.	4·4 sec.	3·3 sec.
50–70 m.p.h.	4·6 sec.	3·5 sec.
60–80 m.p.h.	4·9 sec.	3·5 sec.
70–90 m.p.h.	5·1 sec.	4·0 sec.
80–100 m.p.h.	5·2 sec.	5·3 sec.
90–110 m.p.h.	6·2 sec.	
100–120 m.p.h.	7·5 sec.	

HILL CLIMBING
Max. gradient climbable at steady speed
Top gear .. ..1 in 4·4 (Tapley 500 lb./ton)
3rd gear .. ..1 in 3·4 (Tapley 660 lb./ton)

FUEL CONSUMPTION
Overall Fuel Consumption for 998 miles, 63 gallons, equals 15·8 m.p.g. (17·8 litres/100 km.)
Touring Fuel Consumption (m.p.g. at steady speed midway between 30 m.p.h. and maximum, less 5% allowance for acceleration) 17·48 m.p.g. Fuel tank capacity (maker's figure) 16·6 gallons.

Direct Top Gear
26½ m.p.g. .. at constant 30 m.p.h. on level
27 m.p.g. .. at constant 40 m.p.h. on level
27 m.p.g. .. at constant 50 m.p.h. on level
23½ m.p.g. .. at constant 60 m.p.h. on level
22½ m.p.g. .. at constant 70 m.p.h. on level
20½ m.p.g. .. at constant 80 m.p.h. on level
18 m.p.g. .. at constant 90 m.p.h. on level
16½ m.p.g. .. at constant 100 m.p.h. on level
13½ m.p.g. .. at constant 110 m.p.h. on level
12 m.p.g. .. at constant 120 m.p.h. on level

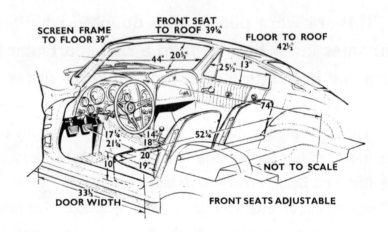

BRAKES
Deceleration and equivalent stopping distance from 30 m.p.h.
0·94 g with 40 lb. pedal pressure .. (32 ft.)
0·53 g with 25 lb. pedal pressure .. (57 ft.)

STEERING
Turning circle between kerbs:
Left .. .. 35 ft.
Right .. .. 34¾ ft.
Turns of steering wheel from lock to lock 3⅔

INSTRUMENTS
Speedometer at 30 m.p.h. .. .. accurate
Speedometer at 60 m.p.h. .. .. 2½% slow
Speedometer at 90 m.p.h. .. .. 2½% slow
Speedometer at 120 m.p.h. .. .. 3% slow
Distance recorder .. .. .. 3% slow

WEIGHT
Kerb weight (unladen, but with oil, coolant and fuel for approximately 50 miles) .. 28¾ cwt.
Front/rear distribution of kerb weight 47/53
Weight laden as tested .. .. .. 32½ cwt.

Specification

ENGINE
Cylinders V-8
Bore .. .. 102 mm.
Stroke .. .. 82·6 mm.
Cubic capacity.. .. .. 5,360 c.c.
Piston area .. .. .. 101 sq. in
Valves .. .. .. pushrod o.h.v.
Compression ratio .. .. .. 10·5/1
Rochester fuel injection
Fuel pump .. .. .. AC mechanical
Ignition timing control Centrifugal & vacuum
Oil filter .. .. AC full flow
Maximum power (gross) .. 360 b.h.p.
at .. .. 6,000 r.p.m.
Maximum torque (gross) .. 352 lb. ft.
at .. .. 4,000 r.p.m.
Piston speed at maximum b.h.p. 3,260 ft./min.

TRANSMISSION
Clutch .. .. s.d.p.
Top gear (s/m) .. .. .. 3·70
3rd gear (s/m) .. .. .. 4·85
2nd gear (s/m) .. .. .. 6·06
1st gear (s/m) .. .. .. 8·14
Reverse .. .. 8·36
Propeller shaft .. .. .. Open
Final drive .. "Positraction" limited-slip differential
Top gear m.p.h. at 1,000 r.p.m. 22·25
Top gear m.p.h. at 1,000 ft./min. piston speed 41·1

CHASSIS
Brakes   Hydraulic drum-type with vacuum-servo assistance; sintered-iron linings.
Brake dimensions: 11 in. dia. drums with 186 sq. ins. of lining working on 333 sq. ins. rubbed area of drums.
Suspension:
  Front: Independent, coil springs, wishbones and anti-roll bar.
  Rear: Independent, transverse leaf spring, radius arms and universally jointed drive shafts.
Shock absorbers:
  Front: } telescopic
  Rear: }
Steering gear: Recirculating ball with hydraulic damper.
Tyres .. .. .. 6·70-15 front, 7·10-15 rear

Chevrolet Corvette Sting Ray

WHEN the Americans took their first, faltering steps towards a sporting car in the European idiom about 10 years ago, the results were, to say the least, disappointing. Their big V-8 engines had power certainly, but handling was suspect and the cars looked rather clumsy. Now, the picture has changed. In most respects, the Chevrolet Corvette Sting Ray, the latest in the Corvette line and introduced just over a year ago, is the equal of any G.T. car to be found on either side of the Atlantic. It falls down on refinement (which is surprising in a model from a country where the most unpretentious cars are notably refined) and wet-road behaviour, a shortcoming for which the cure might be found in the choice of tyres. The silhouette is good, but some of the styling details both inside and out seem fussy. The car is not free of gimmicks, but the performance needs no flattery, the handling is good, and the brakes superb.

The basis of the car, which has plastic bodywork, is a stout, wide box-section chassis frame with coil spring and wishbone independent front suspension and transverse leaf independent springing at the back. The engine is a 5,360 c.c. pushrod o.h.v. with, on the model tested (which was the most powerful obtainable), Rochester fuel injection equipment producing 360 gross b.h.p. at 6,000 r.p.m. Various transmissions (including automatic) are available, the most popular option being the test car's four-speed all-synchromesh gearbox with a limited-slip differential. Drum brakes are used with vacuum assistance and sintered iron linings, and the steering is of recirculating ball pattern. The formula gives vivid performance which, on dry roads, at any rate, is wholly usable at the expense of a not unreasonable fuel consumption.

Performance and economy

SOMETIMES a reluctant starter in the morning, the Sting Ray's engine is smooth, but not quiet. At low speeds there is some clatter coming, presumably, from the valve gear, but as the revs rise it gets very little worse, the mechanical noise at 100 m.p.h. being little more than at idling speeds. Warming-up occupies a short period and coolant temperature in mild weather stays steady at 180°F in or out of traffic, a viscous fan uncoupling itself at high speeds. Super Premium fuel was used throughout most of the test, and experimenting with a mixture of half Super and half Premium produced no pinking, even with the 10.5:1 compression; the optional 11.25:1 ratio on carburetted cars might tell a different story.

Power is even throughout the speed range, giving acceleration which not only *is* swift, but is *felt* to be swift. The standing quarter-mile in 14.6 sec., and a 0 to 50 time of 5.1 sec. gives some indication of how this fairly big car surges forward faster, in fact, than any other car ever road-tested by *The Motor*. Up to 70 m.p.h., each additional 10 m.p.h. is gained at intervals of roughly one second. Top gear acceleration is equally quick and even, the 90-110 m.p.h. time being only a second longer than that from 20-40 m.p.h. The top speed of 146.6 m.p.h. is reached quickly, as the "Maximile" of only 5 m.p.h. less indicates; it is not achieved by an impossibly high, inflexible top gear ratio.

Apart from some fluffing of the plugs in traffic, which was only apparent at about 3,500 r.p.m. once the open road had been reached, the engine proved quite untemperamental. The plugs used were AC 44, AC 46 being recommended if the car is to be used mostly in towns. Taking the revs past 4,000 quickly cleared the plugs, and thereafter they gave no trouble. The test hill restart on a 1 in 3 gradient was easily accomplished; figures for hill climbing in first and second gears were unobtainable because the torque available defeated our Tapley meter. A little care had to be exercised taking off from rest; allowing the revs to drop even momentarily a fraction

—In Brief—

Price (including "Positraction" differential, fuel injection, metallic brakes, and 4-speed gearbox as tested) £2,750 plus purchase tax £573.9.7 equals £3,323.9.7.	
Capacity	5,360 c.c.
Unladen kerb weight	28¾ cwt.
Acceleration:	
20-40 m.p.h. in top gear	5.2 sec.
0-50 m.p.h. through gears	5.1 sec.
Maximum top gear gradient	1 in 4.4
Maximum speed	146.6 m.p.h.
Overall fuel consumption	15.8 m.p.g.
Touring fuel consumption	17.48 m.p.g.
Gearing: 22.25 m.p.h. in top gear at 1,000 r.p.m.	

(*Right*) The cowled, recessed instruments are placed directly in front of the driver. The clock and heater controls are placed on the central console. Switches behind the gear lever control the test car's optional power-operated windows.

Owing to the low overall height the doors extend into the roof line to make access easy. The division in the back window has been removed on later models.

The clean lines of the Sting Ray (*above*) are spoilt by dummy air outlets and false grilles in the bonnet top. The effect with the head-lamps retracted, however, is pleasing.

STING RAY

under 1,000 r.p.m. meant that the engine would stall. At maximum speed the tachometer needle was at about the rev limit with the 3.7:1 axle ratio and oversize (7.10-15 instead of 6.70-15) tyres on the back. The limit of 6,300 r.p.m. was freely used, giving useful maxima in the gears, 100 m.p.h., for example, being easily exceeded in third.

The comparative economy is a measure of the efficiency of the fuel injection system which is of the Rochester type where the fuel is injected not straight into the combustion chamber but into the inlet tract. A constant speed of 80 m.p.h. is necessary to bring the consumption under 20 m.p.g., and the overall figure of 15.8 m.p.g. is not unreasonable when the high-speed cruising and liberal employment of the acceleration are taken into account.

Transmission

THE clutch is long-travel but light. There is a simple adjustment which enables the travel to be shortened at the expense of a heavier movement which would probably appeal to the competition-minded owner. The gearbox is one of the best we have ever encountered in such a high-performance car. The change is smooth, fast, and astonishingly light, and the synchromesh can only be beaten by the most brutal methods, and then only in the lower two gears. There is light spring-loading in the plane of third and top and a lift-up catch on the short, stubby lever is raised for engagement of reverse which, like first and second, whines a little. First is quite a high ratio and the others only moderately closely-spaced but with torque and power available over such a wide range of engine speed, there is a gear for every occasion. Lever travel is short and between third and second, far from there being three distinct movements, it is possible to shift the lever with almost a straight push or pull. With the short drive shafts, there is little "cushion" in the transmission, which sometimes gives a quiet clonk as the drive takes up and throttle movements in first or second gears must be progressive or traffic driving can be jerky.

Handling and brakes

THE Firestone Super Sports nylon tyres with which the car was equipped are adequate at the recommended pressures for dry main roads. Cornering power is high and adhesion under braking or acceleration good, but in the wet a great deal of care and restraint with the throttle must be exercised. The most disappointing feature of the handling is the steering. This is very light and not unusually low-geared for a European-style G.T. car, but very high-geared by contemporary American standards, at 3⅜ turns from lock to lock. And the turning circle is small at 35 feet, meaning that U-turns can be made on a slightly wider than normal main road. For such a fast car, however, it would be better to have more "feel" of the road.

Electric motors raise and lower the dual headlights which give a good beam for fast night driving.

The action is very lifeless, and with tyres which "ride" up white lines, deflecting the car off course, and handling which is otherwise precise, more sensitive steering is necessary. The system has recirculating ball gearing and a hydraulic damper which perhaps does its job too well.

Normally a mild understeerer, the Sting Ray's attitude on a corner is largely dependent on the throttle opening. On twisty roads this can be exploited so that curves are negotiated with a flick of the steering and a jab of throttle. Taking fast bends, with lots of power turned on, the car holds a steady, accurate line. On tighter radii, the driver can confidently open the throttle until a sudden increase in the normally moderate roll indicates that the limit is being reached. A Positraction limited-slip differential distributes the drive to the independent rear wheels, and taking off from rest at high revs in first one can leave two even black lines on the road. Safe and predictable, with good adhesion, the Sting Ray repays a skilful driver with enjoyable fast cornering but there is a tendency to wander slightly on small undulations or surface irregularities at speed.

In the wet, however, the story is different. Wheelspin limits the usable power in first and second, and restraint must be exercised in third—bouts of wheelspin in the eighties can be alarming. The wheels can be locked rather easily under braking but tyres with better wet grip might effect an improvement. Heavy damping combined with fairly soft springs still allows some wheel hop on bumpy corners and there is also some tyre whine on straights, but little or no cornering squeal.

The brakes are astonishingly light and free from fade or snatching in the wet, although when very hot they pulled to the right and smoked. They are drum type, outboard all round with sintered metallic linings and servo assistance, and an adjusting mechanism which operates when they are applied in reverse. The handbrake held the car on a 1-in-3 gradient.

The fuel injection equipment lies between the cylinder banks of the 5.3 litre V-8. The radiator has a separate header tank near the middle of the compartment.

Comfort and control

THE Sting Ray has a smooth, comfortable ride, a little firm at low speeds, but very level. Large, long-wave bumps are taken at speed with a gentle, pitch-free rise and fall. On acceleration the back of the car drops appreciably, as though the rear wheels were "digging in," but there is very little nose-dive under heavy braking. The seats are comfortable and hold the occupants without too much sideways movement on corners. The driving seat has insufficient rearwards adjustment for the tall who like to get away from the wheel, and some people thought that the seat back was too upright. Wind noise is low, even at high speeds, but there is quite a lot of hard mechanical noise and the gear lever rattled a good deal.

The heating system gives a strong blast of hot air through ducts in the side of the central console, and later models have an improved ventilation system bringing the dummy air outlets behind the doors into use as extractors. The car is draught- and water-proof, but visibility is impaired by thick windscreen pillars which are, however, set reasonably far back out of the driver's line of sight. The view rearwards is very poor and the driving mirror shows little of the road behind; the supplementary door-mounted mirror is all the more useful because of the other's inadequacy. Reversing is awkward on account of the poor view aft. Late-series Sting Rays have no division in the back window which should improve matters a little.

Gear lever, pedals and the handbrake (which is under the scuttle) are well placed, and the headlights, retractable by a small switch under the facia, give an excellent beam for fast night driving. The lights take some 12 sec. to reach the raised position, making them valueless for daytime signalling on fast roads unless left up (just when their increased wind resistance is of maximum disadvantage), and there is no separate headlamp flasher.

Fittings and furniture

THE facia, with deeply recessed dials in an arc ahead of him, gives the driver a comprehensive picture of engine and road speed, battery charge, fuel contents, coolant temperature and oil pressure. In the central console there is a not-very-accurate clock and the optional station-seeking radio. The trim is rather garish with a thin-rimmed steering wheel finished in crackled plastic rather clammy to hold. There are no map pockets, but a big, lidded compartment on the passenger's side. The luggage compartment is a carpeted area behind the seats.

The World Copyright of this article and illustrations is strictly reserved © Temple Press Limited, 1963

The rear compartment is fully carpeted for luggage. The seat backs fold downward for access.

Coachwork and Equipment

Starting handle	None
Battery mounting	Under bonnet
Jack	Scissor-pattern
Jacking points	Under body sides
Standard tool kit	Jack, wheel wrench
Exterior lights:	Retractable dual headlights, two sidelights/flashers, two stop/tail, two rear flashers, reversing light.
Direction indicators	Self-cancelling flashers
Windscreen wipers:	Electric, self-parking, two-speed
Windscreen washers	Electric
Sun visors	Two
Instruments:	Speedometer with decimal trip and total mileage recorders, tachometer, water temperature, battery charge, oil pressure gauges.
Warning lights:	Direction indicators, ignition, hand brake.
Locks: With ignition key	Ignition
With other keys	Both doors
Glove lockers	One in facia
Map pockets	None
Parcel shelves	None
Ashtrays	One on tunnel
Cigar lighters	One in facia
Interior lights:	One in roof, two under facia, one in glove compartment with courtesy switches.
Interior heater	Fresh-air type standard
Car radio	Optional
Extras available:	"Positraction" differential; three-speed, four-speed, or fully automatic transmissions. Various engine options.
Upholstery material	Plastic
Floor covering	Carpet
Alternative body styles	Open two-seater

Maintenance

Sump	8 pints, S.A.E. 10/30
Gearbox	2 pints, S.A.E. 90
Rear axle	4 pints, S.A.E. 90
Steering gear lubricant	S.A.E. 90
Cooling system capacity:	27 pints (three drain taps)
Chassis lubrication:	By grease gun every 6,000 miles to nine points.
Ignition timing	4° to 8° b.t.d.c. at 700 r.p.m.
Contact breaker gap	0.019 in.
Sparking plug type	AC44 or AC46
Sparking plug gap	0.035 in.
Tappet clearances (hot):	
Inlet	0.008 in.
Exhaust	0.018 in.
Front wheel toe-in	⅛ in.
Camber angle	+½°
Castor angle	2°
Tyre pressures:	Low speed / Sustained high speeds
Front	24 lb. / 36 lb.
Rear	24 lb. / 36 lb.
Brake fluid	GM Super II
Battery type and capacity	12-volt, 61 amp./hr.

Sting Ray Automatic

Docile but no fossil, and agile but not fragile

 REFERENCE TO THE Sting Ray usually brings to mind visions of a 375-bhp engine, fuel injection, a 4-speed transmission, and all the accoutrements of a competition sports car. However, this car can be obtained in much more docile forms which make it an ideal fast tourer for those who are not interested in the ultimate in performance. With this in mind, we selected our test car with the 300-bhp version of Chevrolet's 327 engine coupled to an automatic transmission, and found that this "power team," as Detroit puts it, was remarkably lively even if some of the keen edge of the really hot version was lost.

It is 10 years since the Corvette was first introduced, with the Sting Ray model following on in 1963, and the car has built up a remarkably enthusiastic following in that time—even comparable to that of the Porsche. In fact, we were surprised to find that some Corvette drivers still wave to each other when passing, even if they do have a tendency to "whomp" their engines at traffic signals. However, it is definitely a most impressive car in many ways and seems to command a lot of attention and respect on the road.

The body design of the 1964 Sting Ray remains little changed from 1963, although we were pleased to note that the central division in the rear window has been eliminated, with a consequent improvement in vision. The basic lines of the car are most original and extremely clean, but the hand of the stylist has been laid on it at some time so that a rather cluttered appearance results at some angles. Because the car is a "Detroit" product (Yes, we know it's assembled in St. Louis), one suspects immediately that the disappearing headlights are some sort of a sales gimmick, but they definitely do improve the aerodynamics of the car and also help to clean up the general appearance of the front end when retracted.

There was some criticism of the quality of the fiberglass body when the Sting Ray first appeared, but we were unable

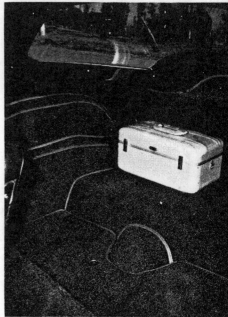

to find any ripples or other faults on our test car. Furthermore, the other characteristics normally associated with fiberglass, such as a tendency toward drumming and magnification of noise, were not present at all, although an occasional creak was audible from somewhere around the rear window when running on uneven ground.

The interior is very pleasing and well finished, and is laid out in the best sports car tradition. Bucket seats are provided for driver and passenger and there is considerable luggage space behind the seats. However, access to this space is through the doors alone, which is extremely awkward when bulky items are concerned, and if the rear window could be hinged in Aston Martin style it would be a great improvement. Another criticism concerns the space allotted to the driver, which is surprisingly small when one considers the overall dimensions of the car.

Applying the standards we used in a recent article on the seating package in various sports cars ("The 99th Man," Jan. 1963) the Sting Ray didn't measure up well at all. Compared to the "average" sports car, the Sting Ray scored 68% and 58% (by the standard for the average and big man), while the average was 80% and 57%. However, provision is made for adjusting the steering wheel, although this cannot be done from the driving seat because it requires the application of a wrench to the steering column under the hood.

In keeping with the other automatic features, the car was equipped with power steering, which we believe to be almost unnecessary on a Sting Ray, but was one of the best we have ever encountered. It was not noticeable at speed, but gave considerable assistance when parking. With an overall ratio of 17.6:1 and 2.92 turns lock-to-lock, the steering was fast, responsive and effortless, and well matched to the diameter of the wood-rimmed, 3-spoke steering wheel.

One of the most interesting features of the car is the independent rear suspension, which is a very successful layout and also a big step forward for General Motors. The system is unusual because it employs a transverse leaf spring, but this is apparently a matter of convenience because there is no room for coils. The two open double-jointed driveshafts make up the top member, and of course these are not splined. Two parallel lower arms pivot low down at the hub carriers and at the differential, and braking torque is transferred to the frame by radius arms.

The result leaves little to be desired because there are none of those qualities, such as a tendency toward rear wheel steering, which one normally associates with independent rear suspension. However, one is aware that the car does not have a solid axle because a slight walking motion is evident under certain conditions of hard driving, but it is never sufficient to be disconcerting and the handling of the car remains completely predictable.

Sting Ray Automatic
AT A GLANCE...

Price as tested	$5016
Engine, cyl/cc/bhp	V-8/5356/300
Curb weight, lb	3050
Top speed, mph	130
Acceleration, 0-60 mph, sec	8.0
Passing test, 50-70 mph, sec	6.0
Overall fuel consumption, mpg	14.8

Sting Ray Automatic

The suspension as a whole is stiffer than we expected but not uncomfortable and does a good job on very bad road surfaces. It is helped considerably by a weight distribution which has a slight rearward bias and this weight distribution has also done a lot for the steering, to the extent that we feel it makes power steering unnecessary under most circumstances.

On the road, the Sting Ray has a giant's stride due to its 300 bhp and 360 ft/lbs of torque at 3200 rpm. The automatic transmission is the Powerglide which, with its two speeds, is certainly not among the most effective transmissions on the market. However, this shortcoming is not particularly noticeable in the Sting Ray because there is power to spare throughout the speed range and the acceleration is exceptional whether the transmission is in high or low.

As far as acceleration is concerned, we were able to draw an interesting comparison with a stick shift Pontiac GTO which we sampled at the same time (see page 32). The Sting Ray tended to squat down on its rear suspension when leaving the line, with never so much as a chirp from its tires, and then gobble up the strip in 15.2 seconds after shifting itself smoothly into high range at 56 mph. In complete contrast, the GTO didn't want to leave the line at all, preferring to sit there burning rubber due to poor weight distribution and lack of an up-to-date rear suspension.

To match the performance of the car, the brakes are adequate for normal fast driving but they will definitely fade and become uneven when used to the limit. When one considers both the weight and speed of the Sting Ray, it would appear to be an excellent car for a disc brake system, and it is surprising that General Motors has not yet adopted discs for this model.

By offering an automatic version of the Sting Ray, General Motors has considerably increased its market for the car. This version is definitely not for the purists, but it is an excellent compromise for those families in which the little woman does her share of the driving, and it has decided advantages for anyone who does the majority of his driving in heavy traffic. With a price tag closer to $5000 than $4000, it is by no means cheap but, on the other hand, it still represents remarkably good value for money when one considers the performance combined with comfort, and the generally high standard of quality throughout the car.

ROAD TEST
Sting Ray

SCALE: 10" DIVISIONS

PRICE
List, FOB St. Louis........$4252
As tested, West Coast......$5016

ENGINE
Engine, no. cyl, type........V-8
Bore x stroke, in......4.00 x 3.25
Displacement, cc..........5356
   Equivalent cu in........326.7
Compression ratio..........10.5
Bhp @ rpm..........300 @ 5000
   Equivalent mph..........118
Torque @ rpm, lb-ft...360 @ 3200
   Equivalent mph...........75
Carburetor, no., make....1 Carter
   No. barrels...............4
   Diameter:.........primary, 1.56
             secondary, 1.69
Type fuel required......Premium

DRIVE TRAIN
Automatic transmission:
  2-speed Powerglide
Gear ratios: 2nd (1.00)......3.36
   1st (1.76).................5.91
   1st (1.76 x 2.10).........12.4
Differential ratio........3.36:1
Optional ratios..........6 ratios
    from 3.08 to 4.56:1

CHASSIS & SUSPENSION
Frame type: Full length box-
  section ladder
Brake type..................drum
  Swept area, sq in..........328
Tire size...............6.70x15
  Wheel revs/mi..............760
Steering type....recirculating ball
  Overall ratio..............17.6
  Turns, lock to lock.........2.9
  Turning circle, ft.........39.4
Front suspension: Independent
  with coil springs, stabilizer bar.
Rear suspension: Independent with
  fixed differential, lateral leaf
  spring & struts, U-jointed drive
  shafts, tube shocks.

ACCOMMODATION
Normal capacity, persons........2
Hip room, front, in........2x19.5
Head room, front..............38
Seat back adjustment, deg......0
Entrance height, in...........34
Step-over height............15.5
Floor height..................8
Door width.................31.5

GENERAL
Curb weight, lb.............3050
Weight distribution
  with driver, %............48/52
Wheelbase, in................98.0
Track, front/rear.......56.3/57.0
Overall length, in..........175.3
Width........................69.6
Height......................49.8
Frontal area, sq. ft.........19.3
Ground clearance, in..........5.0
Overhang, front..............32.0
  Rear......................45.3
Departure angle, no load, deg...17
Usable trunk space, cu. ft....10.5
Fuel tank capacity, gal......20.0

INSTRUMENTATION
Instruments: 160-mph speedometer, 7000-rpm tachometer, fuel
  gauge, ammeter, oil-pressure
  gauge, water temperature indicator, clock.
Warning lamps: parking brake,
  lights, high beam, turn indicator.

MISCELLANEOUS
Body styles available: Coupe (as
  tested), convertible and removable hardtop convertible.

EQUIPMENT
Included in list price: 250-bhp
  engine, 3-speed transmission,
  tachometer.
Available at extra cost: 300-bhp
  engine, automatic transmission,
  limited-slip differential, power
  steering, power brakes, radio.
(Note: many other options available
  too numerous to list.)

CALCULATED DATA
Lb/hp (test wt).............11.3
Cu ft/ton mi...............142.5
Mph/1000 rpm (4th).........23.5
Engine revs/mi.............2555
Piston travel, ft/mi........1385
Rpm @ 2500 ft/min.........4612
  Equivalent mph............108
R&T wear index............35.4

MAINTENANCE
Crankcase capacity, qt.........4
  Change interval, mi.......6000
Oil filter type..............paper
  Change interval, mi.......6000
Lubrication interval, mi....12,000
Tire pressures, f/r, psi.....24/24

ROAD TEST RESULTS

ACCELERATION
0-30 mph, sec................3.2
0-40 mph.....................4.4
0-50 mph.....................6.1
0-60 mph.....................8.0
0-70 mph....................10.5
0-80 mph....................13.2
0-100 mph...................20.2
Passing test, 50-70 mph......6.0
Standing ¼ mi, sec..........15.2
  Speed at end, mph..........85

BRAKE TESTS
Max deceleration, ft/sec/sec...20
  2nd stop...................20

FUEL CONSUMPTION
Normal range, mpg........12–16
Cruising range, mi.......240–320

TOP SPEEDS
High gear (2nd),.........mph 130
1st..........................74

GRADABILITY
(Tapley data)

High gear, max gradient, %....22
1st gear.................off scale
Total drag at 60 mph, lb....120

SPEEDOMETER ERROR
30 mph indicated......actual 30.0
40 mph......................40.0
60 mph......................59.7
80 mph......................79.0
100 mph.....................98.5

ACCELERATION & COASTING

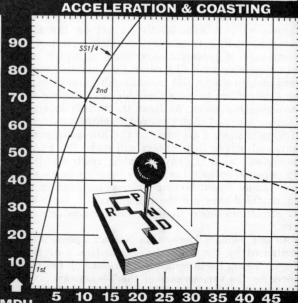

SCG tests two of the cars that caused such a sensation at Nassau!

CORVETTE GS   SCARAB-CHEVY

BY BERNARD CAHIER

FOR EUROPEANS WHO NO LONGER SEE THE SPECTACULAR TYPE OF SPORTS CAR RACING STILL HELD IN AMERICA TODAY, the series of races in Nassau during Speed Week was a wonderful experience and this writer came back home very enthusiastic about what he had seen there. What a change from the rather dull sports car racing we are forced to watch in Europe, thanks to the extremely limiting FIA rulings regarding the type of cars. There is no longer any thundering machinery, outside of the rear-engined Ferrari, which is probably why there are so few sports car races in Europe except for the Championship races such as the Targa Florio, Nurburgring and Le Mans.

Even at Le Mans, the most publicized sports car race of the year, the field was a bit disappointing in 1963 as, against the Ferraris, the opposition was made up of only a few lightweight Jaguars, Aston Martin GTs, one Lola and a lone 5-liter Maserati which was actually the most spectacular car in the race — but it was alone and lasted only two hours. What a race it could have been if a lot of American engined cars like the ones seen at Nassau had been there! But, voila, most of those cars are inelgible according to FIA rulings, and so, logically, most of the American teams did not make the necessary and costly effort to modify their cars for one or two European races; at least they have not done so for the past several years.

With Ford now officially in sports car racing, and GM playing with it (unofficially) a bit on the side, the European racing scene will be much enlivened by the appearance of powerful cars, some of which are now built to comply with FIA rules. I deliberately mentioned Ford and, in a way, GM because only companies such as these can afford to make their big cars eligible for FIA Championship races. At least the sounds and the fascinating sight of high powered machinery will be seen, with enthusiasm I am sure, by the European public this year, and maybe their cheers will make the FIA people understand what the paying public really wants to see. My personal opinion has always been set on this matter and Nassau Speed Week convinced me even more, that big modified cars are the thing, even if some of them are sometimes slower than our smaller production sports cars.

As our readers know, this Nassau Speed Week was dominated by the Chevrolet engined cars of the John Mecom Team. I was certainly impressed by the successes scored by the various cars of the Texas team, and when the very friendly John Mecom asked me if I would be interested in driving some of his cars after the race I was certainly not long in accepting this offer. Monday morning after the last race, with the sun still shining brightly, I went to the hangar near the track to take temporary possession of whatever was there to be driven. I was pleasantly surprised when I was told by the mechanics that I could drive one of the lightweight Corvettes, the Scarab-Chevrolet which had won the two major races, and, if they could get it ready on time, the Lola Chevrolet which has also won two races earlier in the week, driven by Augie Pabst.

All of this was handled in a very informal manner, and after quickly checking the oil, water and gasoline I took the driver's seat of Corvette No. 65, which had taken eighth overall driven by John Cannon. Two of these Corvettes had been entered in the main race, the other one finishing fourth overall driven by Dick Thompson. Actually these two cars would have finished third and fourth if it had not been for their faulty hoods, which kept opening all the time and forced these cars into the pits many times, since the tape used to secure the hoods did not last long. I personally thought that these two cars put on a tremendous show and that their performances can be explained by the fact that they were actually much different from the standard Corvettes which had such a hard time against the Cobras last year. They have a different chassis, altered suspension, disc brakes all around. Their weight has been cut down from about 3,000 pounds to 2,000 pounds, and their power plants were no longer the 327 cubic inch engine but a 377 (6.2 liters) version giving no less than 480 horsepower, with a rather formidable torque of 500 lb/ft at 4,000 rpm. Inside, the cockpit of the car had been stripped of any unnecessary weight, and lighter but very comfortable bucket type seats had been installed. The windows were made of plastic and could be adjusted by a strap to any position, almost a luxury in such a car. The driving position was excellent, the pedals well placed, but the sight of the enormous hood covering the mighty V-8 engine, fed by four huge double throat Weber carburetors, was very impressive!

Even though the engine had not been fired since the day before, it came to life immediately, filling the hangar, *and* the cockpit, with a deep powerful sound. Soon I was rolling out, and as I was warming it up around the track (and actually looking for the way to go, as I had not yet been around the circuit) my first impression was of how docile the car was at low speed, how light and direct the steering was in spite of the enormous Goodyear tires fitted on the car. The car was big all right, but did not feel it. For a competition machine the brakes, clutch, throttle and gear box were particularly pleasant to use, in a way like a refined GT car, and I was hardly expecting that.

Cobra-hunting, the Sting Ray squats under fierce acceleration.

Stock-car Goodyears and alloy wheels provide maximum bite.

Interior of the coupe has a very production-like appearance.

Oil coolers rectified overheat that first burned differential.

Four DCOE-58 Webers on special tuned-length manifolding help obtain what is reported at well over 450 hp from the engine.

CORVETTE GS and SCARAB CHEVY

After a couple of cool laps I started to feel at home at the wheel of this car and began to gradually put on the steam. Gone was the docile Corvette; the lamb turned quickly into a tiger and I was breathless with the stupendous acceleration of the car. With 480 horsepower for a weight of some 2,000 pounds, the Corvette Grand Sport is not the sort of car you drive very often, and at first I was concentrating mainly on applying the right amount of power in order not to spin the wheels wildly or come out of corners sideways, which, believe me, is very easy to do. If you really want to show off you can leave rubber marks for a quarter of a mile, as you still get wheel spin when putting it into third gear. Once accustomed to this immense power, and reserve of power, the car is no problem to drive fast — which is still not really racing speed! The handling seemed indeed free of any major vices and the car was equally at ease in slow, medium or fast bends. However, I noticed that at high speeds on the straight the combination of power, acceleration, and the not-too-good aerodynamics of the car (it has an enormous frontal area) make the front of the car lift and the steering becomes light indeed. Also, on fast bends, like the ones at Nassau leading to the pits and under the footbridge, the car was not too happy with the very rough surfacing of the track and was wandering enough to keep me fully alert. The handling characteristic of the car was quite neutral, but I felt at times that it oversteered some, perhaps because I applied power too hard. The ride was rather comfortable and the seat gave good support, but the noise inside and out was deafening, especially with the windows open, since the exhausts come out right under them. I wished I had taken ear plugs as it took me several hours to recover my hearing completely!

What really impressed me with this Corvette, besides its tremendous power and excellent handling, was the lightness of the various controls and really the suppleness of the engine which would go down to 1500 rpm in top gear without protestations. Granted, 377 is a lot of cubic inches, 480 is a lot of horses, but I am quite sure that the car could be started in top gear without too much fuss; that is certainly not the sort of thing you can do with a Ferrari, where you simply must keep those revs up to get the power and supple-

Much-munched Scarab survived many off-course excursions and went on to win Nassau main.

Ample cooling and rugged components added to the reliability that helped out-run opposition.

Traco-built 327-inch Chevy pumps out 425 hp., also uses four Webers, tuned intake system.

42

CORVETTE & SCARAB

ness. Another thing which surprised me was how nicely the engine would idle, and this without apparently upsetting the plugs. Needless to say, no acceleration times could be taken, but I understand that this Grand Sport Corvette will do zero to 100 mph in nine seconds flat and the standing quarter of a mile should be under 12 seconds.

Driving the Grand Sport Corvette for over an hour was an exhilarating but all too brief experience to really discover all of its qualities and faults. But still it was enough to make me half deaf, and to understand better what the drivers who raced those cars had told me about them. Their comments were highly favorable, particularly those who drove these cars for the first time, like Jim Hall, Penske and Cannon. It is a very large car but quite easy to drive, immensely powerful but very controllable, and having general performance and handling which put it well ahead of the Ferrari GTO and Ford Cobra. Granted, the Corvette Grand Sport is not homologated as a true GT car (and it is a pity), but it is a true GT machine in every sense of the word. It is exactly the type of car which the touchy organizers of a race like Le Mans would like to see there, because it is in a sense so near to a genuine high performing GT car.

Stepping from the Corvette Grand Sport into the Scarab was, of course, quite a change. Here was a pure sports-racing or modified car, weighing a good 500 pounds less than the lightweight Corvette, but nevertheless having 425 horsepower, a power-to-weight ratio substantially better than the mighty Corvette, and that meant more thrills and experience for me. Designed for Lance Reventlow by Eddie Miller, Jr., and built by Phil Remington, the rear engined Scarab is a sort of classic sports-racing car built with the sturdiness of a Cooper Monaco, but with that unmatched American finish; all the material used is of aerospace quality. The car was at first equipped with an aluminum Buick engine, then an Oldsmobile engine was fitted into it, with which A. J. Foyt scored a worthy second place at Laguna Seca last October. After that this car was turned over to John Mecom and the Oldsmobile engine was replaced by a special 327 cubic inch Corvette engine whose preparation was given to Traco Engineers of California. The result was most convincing. This engine, fitted with four double throat Weber carburetors was giving a healthy 425 hp, with a torque of over 400 lb/ft at 3,500 rpm, and A. J. Foyt won the two main races at Nassau with this perfect combination of car, driver and engine.

My drive of the Scarab was arranged just as informally as had been that of the Corvette. We checked the oil and water, added a little gas, and the car was rolled out of the hangar to be cleaned a little, as it was rather filthy after more than 250 miles of racing and a cross country trip by Foyt in the latter part of the race. The cockpit was, in fact, full of sand and grass, but that did not matter much as long as I could see through the windshield, and I could. The cockpit of the car itself is very roomy by today's sports-racing car standards and there is no problem getting in. My main problem was to find a possible driving position, since the seats could not be adjusted and the pedals were awfully far away for me, Foyt being a rather big fellow. Anyway, by sliding down to use clutch and brakes I could manage, and I was soon off on a now more familiar track. As with the Corvette, the engine started immediately, and although it did not feel as smooth, I thought that for a pure racing powerplant it was remarkably tractable, and the use of low revs while I was warming it up around the track did not seem to affect its behavior. There were all kinds of grinding noises in the cockpit, but the exhaust being behind me it was not nearly as bothersome as on the Corvette. The steering was beautiful but the pedals were always a bit of a problem, and the gear box (operated by a short little lever) was not as refined as that of the Corvette. It was still a good box and the only difficulty I occasionally had was going down from third to second, perhaps because the gates had been a bit bent during that heated race. This gear box, incidentally, was a Colotti Type 37, especially made for American engines. It is said to be very strong, as shown by the several cars which used it at Nassau, and by Clark and Gurney who were also using it in their Indianapolis cars.

After my two warm up and reconnaissance laps I started to have a little go with the Scarab and was immediately stunned by the acceleration of this 1500 pound car pushed by 425 horsepower. With such a car it became even more necessary to be careful as, for example, coming out of the straight those shut-off markers were flashing by awfully fast with that next turn coming uncomfortably quickly if you were not prepared for it! In corners the handling of the car was superb, and although a rear engined car, the Scarab, under normal driving conditions like mine, felt more understeering than oversteering, but it was generally quite neutral. The brakes were very powerful and efficient, and apparently did not seem to have suffered from 250 miles of hard racing. Visibility through the windshield made of Plexiglass was satisfactory, and I was surprised that my cap was not blown off once, meaning that there was little wind turbulence in the cockpit.

The more I went around the track the more I liked the car, and only the prospect of running out of gas made me stop after an hour of the most exciting type of driving I have ever had. By the end I was motoring along quite quickly, but always treating the car with the utmost respect, and although I was alone I did not dare to try to prove anything, even to myself. Still, I was going around very comfortably in 3 minutes 15 seconds, which is no fast time for this car which has been taken around in 2 minutes 41 seconds, but still fast enough to give you that certain feeling found in this exceptional type of car. Compared with the Corvette, the Scarab felt naturally more agile in corners, its power was better distributed to the road, and it felt very good on fast turns even on the rough sections of the track. Besides being exciting, the car was actually fun to drive and it was wonderful playing around corners with it, its behavior was so healthy and frank, and WOW! – that acceleration when applying the power coming out of those turns! You really felt your insides going up and down. Dirty and more deaf than ever, I stepped out of the Scarab with a big grin on my face. I had driven one of those real brute power sports racing cars which look so spectacular on the track, and this experience had been an unforgettable one. I thought with great respect of the people who built it, prepared it and drove it.

By the time I got back to the hangar I noticed the half sorry and half glad faces of the mechanics who told me that unfortunately I could not drive the Lola Chevrolet, not this time, anyway. It seems that when they were preparing the car they started the engine from the outside, not realizing that the car was in gear. So off went the Lola in the hangar all by itself with all the mechanics trying to stop it. Well, with all that power under the hood this is not easy, nor is it an easy task to get into the car! Fortunately a large station wagon ended the Lola's lonely drive after a fifty foot trip, and fortunately the damages were limited to a broken headlight for the Lola and a small dent for the station wagon. The car, however, could not be driven after that, as the gear box did not function properly, which may explain why it was in gear when they thought it was in neutral. Apparently its insides were all mixed up. Regretfully I left the Lola and this wonderful group of mechanics of the John Mecom Team, who had made my day memorable by allowing me to drive two unusual, exciting and high-powered machines.

"G.P. race? I thought they said Gee Bee race!"

PATRICK McNALLY tests the

CHEVROLET CORVETTE STING-RAY

THREE-HUNDRED-AND-SIXTY brake horse power, an independent rear end and glassfibre body—these are the ingredients which make the Chevrolet Sting-Ray one of the fastest sports cars available today. AUTOSPORT has already tested the open version of this model, but there are several differences which should be mentioned.

The subject of our test was the "fastback", or closed version, which with fuel injection, produces well over 200 b.h.p. per ton. The main differences between this car and the one we had tested earlier were that it was slightly heavier and the gearing was a little altered. The latter was caused by over-sized rear tyres which were fitted (7.60 x 15).

The 5,340 c.c. power unit is the well-tried Chevrolet V8 which, with Rochester fuel injection, hot cams, light alloy heads, a compression ratio of 11.25:1, produces its advertised 360 horses without any trouble. The torque figure is very impressive, 352 lb./ft. being produced at 4,000 r.p.m.

Coupled to this compact engine is a Borg-Warner four-speed all-synchromesh gearbox. The ratios are: Top, 3.7, 3rd, 4.85; 2nd, 6.06; 1st, 8.15. These are overall ratios, with the 3.7 final drive. The clutch is a 10 in. single dry-plate unit hydraulically operated and centrifugally assisted. The final drive is by hypoid bevel with a limited slip differential.

The chassis is a ladder-type box-section frame with glass reinforced plastic body. The glassfibre body is unstressed and is attached to the chassis on 12 main mounting points. The front suspension is by double wishbones with coil springs, telescopic dampers and an anti-roll bar. The rear is independent with single lower wishbones and fixed length drive shafts, with a transverse leaf spring and telescopic dampers.

The brakes have a total swept area of 334 sq. in.; the fronts are 11.2 in. in diameter, as are the rear. The front shoes are 2.74 in. wide as compared with the rears, which are 2 in. They have twin master cylinders which are vacuum servo-assisted, hydraulically operated and have self-adjusting cintered linings.

The steering is of the recirculatory ball type with 3.4 turns from lock to

lock. The turning circle is 40 ft.

Our car was fitted with cast aluminium centre lock wheels with 6 in. rims. These were shod with Firestone Super Sport 7.60 x 15 in. tyres on the rear, with 7.10 x 15 in. on the front.

On the road the V8 showed itself to be remarkably flexible and not in the least bit fussy. In the early stages of the test it had been inclined to stall in traffic, but this was quickly rectified by tightening up the slow-running screw—the cause of the trouble. Once done, we had no trouble and the car would run from one side of London to the other without any of the temperament of a high performance G.T. car. It should be stated, however, that the clutch caused one or two anxious moments, for although it was not fierce in operation it was necessary to have at least 1,500 to 2,000 revs. in hand if one wanted to move forward in traffic. There appeared to be absolutely nothing below 1,000 r.p.m., and if the clutch was let in as one would with most 5-litre cars, to slip gently forward in slow-moving traffic, the resulting stall would necessitate the ignominy of trying to restart the hot engine, and this was not always easy.

The engine was extremely flexible above 1,000 r.p.m. and it was possible to cruise as low as 30 m.p.h. in top gear. Below this figure there was some transmission snatch, which was not surprising in the circumstances.

First gear gave a maximum of 65 m.p.h. which greatly influenced the 0-60 time. This was exactly 6 secs.; 50 m.p.h. was reached in 4.9 secs. and 30 in 2.85 secs. Second gear gave 86 m.p.h., which made the 0-80 figure very good indeed. The mean time for this figure was 9.85. 111 m.p.h. was the maximum achieved in third, while the top speed of this car was 148 m.p.h.

It is interesting to compare these figures with those of the soft-top version, for they show the lower gearing of the latter car but the better streamlining of the fast-back. The 0-80 of the soft top was 9.5 secs. and the 0-60 5.4 secs., but the 0-100 is almost identical, the open version recording 14.2 secs. and the closed version 14.25 secs. The standing quarter mile was covered in 14.5 secs.

The power is readily available all the way through the rev. range, but maximum is found about 4,000 r.p.m. There is little point using over 6,000 r.p.m. although the unit will rev. to 6,800 and 7,000.

In our previous road test the brakes were heavily criticized, for they did not provide the driver with any confidence. When hot they tended to pull, and wet weather affected their operation also. Despite the fact that, in theory, exactly the same brakes were fitted, it was interesting to see how much they had been improved. The pedal pressure was lighter and, although the brakes still tended to pull, they certainly imparted far more confidence than before.

The steering, although light in operation, was too low geared for a car of this performance where, in my opinion, the maximum turns from lock to lock should be 2½. The high-speed straight-line running characteristics were good, but could definitely be improved by high-geared steering.

The gearbox is superb. Of all the high performance cars I have driven, it has the slickest and lightest change. It is interesting to note on a graph the time taken between gear changes when the graphs of the E-type, Aston Martin and Chevrolet are compared. It is not just brute force which makes the Sting-Ray's figures so impressive, but also the speed of the gear change.

As road transport the Sting-Ray has tolerable manners. Although inclined to pitch and roll by European standards, all four wheels seem to remain on the ground; it is possible, when really pressing on, to lift a front wheel. The rear suspension copes admirably with the power, and fast take-offs are not difficult. Care must be taken in the wet, however, for imprudent use of the throttle can easily cause fascinating diversion.

The seating is most comfortable, the gear lever comes nicely to hand and the adjustable steering wheel can be ideally positioned. The instrumentation is as one expects, with rev. counter and speedometer in pride of place directly in front of the driver flanked by temperature gauges, etc. One of the

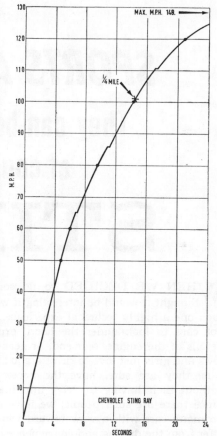

ACCELERATION GRAPH

nicest features of the test car was the electrically operated windows—an unusual luxury for a high-performance G.T. car.

In summarizing, it is fair to say this is a very exciting car to own. The performance in a straight line is tremendous, the appearance not displeasing and the effect on the fair sex rewarding. If you can afford just over £3,000 and don't want to meet too many identical cars on your journeys (and are not too enthusiastic about ultimate road holding and braking) this may well be the car for you.

SPECIFICATION AND PERFORMANCE DATA
Car Tested: Chevrolet Corvette Sting-Ray two-door G.T., price £3,432, including P.T.
Engine: Eight cylinders (two fours in V), 5,340 c.c. Push-rod operated overhead valves in light alloy heads. 360 b.h.p. at 6,000 r.p.m. Compression ratio 11.25 to 1. Rochester fuel injection with automatic choke. Coil and distributor ignition.
Transmission: Single dry plate clutch. Four-speed gearbox with all synchromesh. Ratios 8.14, 6.14, 4.85 and 3.70.
Chassis: Steel chassis with glassfibre body, independent four-wheel suspension by coil springs and double wishbones on the front with anti-roll bar, and transverse leaf spring, trailing arms and torsion bars on the rear. Drum brakes both front and rear, total brake area 328 sq. in. Twin master cylinders and vacuum booster. 7.60 and 7.10 x 15 tyres on bolt-on disc wheels.
Equipment: 12-volt lighting and starting. Rev. counter, speedometer, fuel gauge, petrol gauge, oil pressure gauge, ammeter, windscreen wipers and washers, heating, demisting, flashing direction indicators, radio, clock.
Dimensions: Wheelbase 8 ft. 2 ins. Track (front) 4 ft. 8¼ ins. (rear) 4 ft. 9¼ ins. Overall length 14 ft. 7 ins. Width 5 ft. 9¼ ins. Weight (approx.) 29 cwts.
Performance: Maximum speed 148 m.p.h. Speeds in gears: 1st, 65 m.p.h., 2nd, 86 m.p.h., 3rd, 111 m.p.h. Standing quarter-mile 14.5 secs. Acceleration 0-30 m.p.h., 2.85 secs.; 0-50 m.p.h., 4.9 secs.; 0-60 m.p.h., 6 secs.; 0-80 m.p.h., 9.85 secs.; 0-100 m.p.h., 14.25 secs.; 0-120 m.p.h., 20.75 secs.
Fuel Consumption 15 m.p.g.

SPORTS ARE LIKE SHORTS...
They can be worn by anybody! That of course includes the Corvette.

STING RAY

WHEN WE DECIDED to do some testing, we thought it would be interesting if we did not make each one a highly technical experience. Everyone does not care to look under the hood, explore the super details of the engine, rear end, or gear box. Some fellows and girls just love to drive a good car and certain cars, they are sure, have the necessary ingredients without their having to delve into masses of figures. Since there are such people, we have conducted tests on many cars with a view toward what they have to offer from the driving and convenience of the machines involved. We are convinced that among the best possibilities are the American Sports Cars which offer an excellent choice in 1964.

Too many people think of these Sports Cars as something only for the racing driver, the fellow who wishes to compete on the track or road circuit, and not for anyone else except a young fellow who gets a thrill out of leaving rubber at every stop light, and manages to keep the police busy giving out tickets. This is not a true picture. Perhaps TV and the movies may have had something to do with forming it. We drove these machines, and were amazed at the *safety* of the cars. It is a fact that a fast car must be a safe machine,

Proof the pudding. While he golfs the wife takes the Sting Ray for a shopping spree. Any Corvette is a double duty expert.

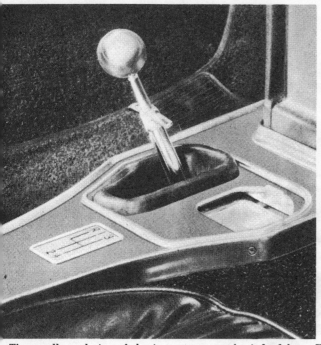

The excellent placing of the instruments can be judged here. Everything is right where you can see it. The large knobs atop the shift lever affords the driver firm grip. Notice the large safety catch which prevents accidental shifting into reverse, a feature.

and they are so designed. Their safety factor is greater than that of any other type of automobile, chassis - wise and control - wise. This should always be kept in mind.

These cars need not always be driven at breakneck speeds. They function equally as well in traffic and on local roads. As a matter of fact parking one of them in a crowded supermarket parking lot proved a lot easier than having the same task with a much larger automobile (whose size actually meant little in most cases) that had all sorts of power equipment. We also discovered that there was usually enough space for packages and other incidentals in most Sports Cars.

The first car we tested was the Chevy Sting Ray, a red Sports Coupe. The one thing we did not like about it was the fact that we had to return it to General Motors. The car was equipped with a four-speed stick shift, power brakes, and windows. There was an excellent Am-Fm radio, a great big heater, which was more than welcome in the 0 degree cold weather, large clear and well-lighted dials, a clock with a second hand that could be seen without hunting all about the dash. We could see how fast we were going, how quickly the engine was turning over, if the generator was charging, for there was a good ammeter, and just how high the oil pressure was. No blinking light, no guesswork.

The windshield wipers were strong enough to quickly remove the heavily falling snow and an adequate defrosting system prevented fogging up of the glass surfaces. The leather upholstered interior and carpeting were tastefully done. In back of the passenger seats we found a very large compartment for luggage, packages, and even room for the young children. Safety belts were provided, and if children are frequent passengers, extra safety belts could be installed.

In checking the gas tank we noticed that the filler cap was large and easily accessible. No fumbling or spilling all over,

The bucket seats most certainly were the most comfortable we ever experienced. They held firmly and gave a full measure of adjustment so that the proper distance from steering wheel and pedals was quickly obtained. This afforded excellent road sight.

The doors, which curve into the roof, fitted flush and accurately with the body sides assuring draught free rides. This also made for easy access. Steering was light, sure and positive. Shifting by hand was quick. The safety release on the shift allowed for no mistakes. The shift knob was round, large and afforded a positive grip.

Normally the hung foot pedals make you feel as though you are pushing through the bottom of the floor rather than the good old feeling of fore and aft motion. Yet the Sting Ray's shoes were fitted better and the slip-off sensation was absent. Clutch motion was easy and sure. Brakes responded excellently. They did not lock or pull. You always had complete control. At no time did the machine 'wallow'. It did not drive you, you drove it. Good control, good seating and certainly big-car comfort was found inside the Covette.

The exterior fiberglass body, with the exception of some seemingly unnecessary depressions, was expertly designed. One improvement we would like to see is true wire wheels with knock-off hubs. Two minor things that could be eliminated were the script at the rear and the somewhat tin-toy appearance of the standard wheel disks.

Color selections were good and the choice was excellent. Although we lean toward the red and black combinations, lighter and more feminine colors are available for the ladies. They can drive these cars, by the way, without having to be a super Amazon and will actually fall in love with them as most wom-

Three views of the newest Corvette Sting Ray convertible. Strong conitnental appearance is evident. Seldom has there been a more dashing automobile. That America has something to offer cannot be denied when one discovers this car. Little is left to be desired.

en do once they have the opportunity to try them.

To sum up our experiences with the Covette Sting Ray we must say that it was a delightful one. We were pleased with the car and found it lives up to all its vaunted capabilities plus the outstanding and much overlooked fact that it is an excellent car for family use. The wife can use it on her daily shopping trips, visiting friends, and the man of the house can use it on the private roads of his Sports Car club. This is one car than can really serve the entire family with the greatest safety measures available today.

The best way to describe the 1964 Sting Ray: A Sports Car with a dual personality. The Sport Coupe, as well as the convertible, feature far better ride qualities, higher horsepower ratings for the two top engines, cleaner styling and more comfort inside.

Improved four-wheel independent suspension, short-wheelbase chassis

and a shift of weight toward the rear has helped in great measure to improve the already excellent ride. Front and rear, new variable-rate springs size up bumps as they take them on. In other words they work softly on small bumps and offer more resistance to large ones. Beside this, all shock absorbers have been recalibrated, and noise has been cut down by a host of innovations designed to reduce and isolate vibration. This results in a firm, smooth, quieter ride on country roads or in town, plus improved open-road handling and performance characteristics.

The belt-line remains unchanged, sweeps around the aerodynamic design. The easy contour of the body panels is underlined by new rocker moldings. New, too is the one-piece rear window through which a lady can see without straining in the seat. At the front, power-operated retractable headlamps go smoothly into the hood. The hood itself is styled with clean, uncluttered surfaces. The Sports Coupe has extended doors that go well into the roof for easy ins and outs. No one has to duck away from this one.

Now a little about the Convertible. Fore and aft, there are trim wraparound bumpers and full length windsplits on the hood and rear deck. There is an aluminum grille, distinctive wheel covers with simulated knock-off hubs and a pleasing absence of chrome and useless decorations. You can select a removeable hardtop, or just leave the soft top packed neatly away for fast easy driving pleasure in the clean air. Perhaps both the hard top and the convertible would best suit the needs of a cold climate. Whichever way, the Corvette promises great times, sparkling performance.

The Ball-Race steering makes for accurate placing, light and easy control. The power teams are a quartet of restless V8's coupled to a selection of versatile transmissions. With all this the Sting Ray lives up to its promises.

Speaking of luxury as well as the other features of this great car, just seat yourself into the cockpit of the automobile. There are plush bucket seats to hold you in place and in comfort. The deep-hub steering wheel with its simulated walnut

This year simulated grills have been removed, giving the hood a smoother appearance. Also new is an interior ventilator system which improves air circulation and adds to passenger comfort. One regrets that Knock-off wire wheels are not offered as standard.

rim feels just right, and is standard. The adjustable steering column, another well thought-out addition to driver comfort, is standard too.

Though the upholstery is not top grain fine leather, the elegant leather-grained vinyl might just out wear anything else. Certainly the color keyed interior (in black, red, dark blue, silver, white or saddle) leaves nothing wanting and spruces up both seats and hooded areas of the instrument panel. The upper part of the doors are pad-pleated with leather-grain vinyl while the lower part is carpet covered. At a slight extra cost you may choose your own color-keyed genuine leather seats.

Once inside you will also notice the extra leg room, not exactly standard in Sports Cars. There is deep twist carpeting beneath your feet, not rubber mat, or plastic sheet. It is warm and comfortable feeling and assures high-heel and scuff protection. Behind you, as we have already said, is a fully carpeted luggage area. For secret out-of-sight stowage, there are two concealed compartments beneath the luggage area floor. One for your tools, the other for valuables.

The bright new door release, shift lever and directional signal knobs are all chrome. The instrument cluster is very functional. There is in the group a speedometer, tachometer, oil pressure, ammeter, fuel and temperature gauges. All are of good size, clear and at night, well lighted. The smart center console does its share of work, having the electric clock, heater controls and an AM-FM transistorized radio that is optional at extra cost. Even the breeze is at your fingertips. Power operated windows are available if you want this bit of convenience at optional cost. The ventiplanes though hand cranked are easy to manage. Additional equipment, like power steering, power brakes, automatic transmission and air conditioning, most, if not all will appeal to the feminine driver, are extras.

Driving in cold weather is a pleasure with the Corvette. It's heater is second to none. It is coordinated with the cowl inlet ventilation system. Warm air comfort and a frost-free windshield are as close at hand as the two control knobs. Something new for the 1964 coupe is the rear compartment ventilation blower. This is mounted in the body side panel behind the drivers seat and designed to draw air out of the passenger compartment via louvered body openings. The blower is actuated by a pull type switch located near the retractable headlamp control and operates only when the heater is off. The headlamps are operated by two reversible electric motors. Just in case, they may be operated manually as well or locked in open position. A cockpit light warns when lights are on when not in fully open position. If you want your lights motor-operated, it will cost a bit more money, but it will save manual handling.

Some of the standard equipment included is a Delcotron generator, hydraulic self-adjusting brakes and long-life exhaust system. Also, seat belts, amber lenses on front parking and turn signal lights, inside and outside mirrors, recessed safety reflectors in the door panels. Padded armrests, sunshades for driver and passenger, electric windshield wipers, push button windshield washer, directional signals, parking brake alarm and courtesy lights are also standard.

The exterior and interior of the Corvette Sting Ray has been discussed. Now for the innards, and there is plenty to talk about and plenty to choose from. All four versions of the V8 engine have a 327 c.i. displacement, 4" bore and 3.25" stroke. The 250 hp standard engine features a single 4-barrel carb, 10.5:1 compression ratio, hydraulic valve lifters and a dual exhaust system. The 300 hp power plant (optional at extra cost) adds a large 4-barrel carburetor and larger intake valves. Both the new 365 horsepower V8 and 375 horsepower Fuel Injection V8 (also optional at extra cost) have a special camshaft, new cylinder heads with extra-large valves, domed aluminum pistons, an 11.0:1 compression ratio, a high speed valve system with special valves, mechanical valve lifters, ribbed aluminum rocker covers and a 5-quart oil capacity (4 quart capacity on the 250 and 300 horsepower engines).

The standard transmission is 3-speed Synchro-Mesh. Ratios 2.58:1 first; 1.48:1 second; 1:1 third; and 2.58:1 reverse. For a variety of engine and rear axle combinations, 4-speed Synchro-Mesh is the answer. Ratios with 250 and 300 horsepower V8's: 2.56:1 first; 1.91 second; 1.48:1 third; 1:1 fourth; and 2.64:1 reverse. Closer ratios with the 365 hp engines: 2.20:1 first; 1.64:1 second; 1.28:1 third; 1:1 fourth; and 2.27:1 reverse. With both 3 and 4 speed Synchro-Mesh, there's a 10" semi-centrifugal diaphram spring clutch with a lighter aluminum 360 degree clutch housing. And the floor mounted shift for the four speed transmissions has a mechanism on the shift lever to prevent unintentional reverse engagement. The automatic Power-glide transmission, available with two engines, offers truly effortless driving. And you can select the posi-traction axle ratio best suited to your choice of power teams.

Here's a rundown on the innovations that make the ride of the Sting Ray softer and quieter and flatter cornering for true Sports Car performance. New variable-rate front coil springs and variable-rate rear spring. The amount of deflection decreases in proportion to the severity of the road shock encountered. Recalibrated front and rear shock absorbers. New rubber jounce bumpers, front and rear, provide a progressive rate of cushioning. Revised exhaust system components with tuned mufflers to reduce exhaust noise. New loop-type tail pipe hanger affords additional vibration insulation and the front exhaust pipe hanger is now attached to the transmission rather than to a frame crossmember to further reduce vibration. New rubber body mounts for both the sports coupe and the convertible are provided. Both shift lever boot and shift lever linkage make use of rubber to further isolate vibration and noise. There is special insulation between carpet and fiber glass panels in the front and rear compartment of the sport coupe body. Also, new insulating blankets on rear of front wheel houses. These reduce noise and transmission of vibrations. These is a new rear end panel stiffener to muffle resonance.

The rear suspension features a three link independent system at each wheel. A radius arm running from the frame to the rear spindle support a control rod attached to the differential and wheel spindle, and the double universal-jointed tubular axle shaft combine to provide nearly vertical wheel move-

Looking directly at you. The clean uncluttered hood is a great improvement over older design. The rear window is now one piece, giving unobstructed rear vision.

ment. One new multi-leaf transverse spring with varied leaf camber is bolted to the differential carrier and extends from rear wheel to rear wheel. It takes only vertical suspension loads. Braking, acceleration and lateral forces are transmited by the radius arms, axles and control rods to the differential and the frame. The differential assembly is attached to the frame by a rubber cushioned bolted-in crossmember. Small movements of the differential are taken up by universal joints on the one piece propeller shaft. For increased torsional rigidity and beam strength, the Corvette Sting Ray convertible has a steel framework virtually surrounding the passenger compartment, while the sports coupe has overhead steel roof members as well.

It is possible to add tire and handling stability with optional wide-rim aluminum wheels. Vented fin design for cooling; knock-off hubs for easy wheel changing.

All engines have independent mechanism for each valve; temperature-controlled fan; precision-machined forged steel crankshaft; premium aluminum main bearings; full-pressure lubrication system; full flow oil filter; automatic choke; and a 12 volt electrical system: Oil-wetted polyurethane air cleaner element. All-aluminum cross-flow radiator. Positive closed-type crankcase ventilation.

The chassis has double-acting freon-bag shock absorbes. The balanced steering linkage with 19.6:1 overall ratio can be reset to 17:1 which is standard for power steering. Hydraulic 11 inch brakes with fade-resistant bonded linings are stock. Black 6.70 x15" tires are standard but you may have as optional extra nylon blackwalls or rayon whitewalls.

The body is fiber glass reinforced plastic with Magic-Mirror acrylic lacquer finish. There are seven colors: Tuxedo Black, Ermine White, Riverside Red, Silver Blue, Saddle Tan and Satin Silver. Three color choices are available with any body colors, for convertible tops, white, black, and beige.

Additional optional equipment includes back-up lights; sintered-metallic brake linings; cast aluminum wheels with 6" rims and knock-off hubs; 36.5 gallon fuel tank can be had with the sports coupe; and off-road exhaust system; full-transistor ignition system, which can only be had on the 365 and 375 horsepower engines. There is a special performance equipment grouping with the 375 hp engine that offers special power brakes with finned drums, built-in cooling fans, front brake air scoops, special sintered-metallic linings, and forward-driving self-adjusting feature. Beside all this there is a dual circuit brake cylinder, a heavy-duty stabilizer bar, heavy-duty front and rear springs and shock absorbers. Quite a choice, indeed.

Generally speaking the dimensions are as follows: Wheelbase 98". Overall length, 175.3". Overall height: Convertible with soft top up, 49.8"; Convertible with hardtop, 49.3". Sport Coupe, 49.8". Cowl height to ground, 34.9". Door opening height to ground: Convertible, 45.6"; Sport Coupe, 46.8". Road clearance, 5.0". Overall width, 69.6". Tread: front, 56.3"; rear 57.0"

PAUL E. HANSEN PHOTOS

CORVETTE vs. COBRA

Pointing Up the Differences, and Similarities, Between Two Fine Road Cars

EVER SINCE THE prototype AC Cobra appeared on the West Coast with entrepreneur Carroll Shelby grinning behind its steering wheel, the verbal battle has been waged. Shortly after Ol' Shel's public introduction of the car, variously comprised teams of competition Cobras appeared on the country's road racing circuits and another sort of battle was begun. The Cobra's arch-rival, of course, was the Chevrolet Corvette, until then the king of the domestic sports car heap.

Although the Cobra, in racing trim at least, has proved the swifter of the two on the road circuit and drag strip, the heated discussions persist between protagonists of each make, who are not willing to concede to the other a single point on which is the best "street" sports car. So, in an attempt to at least organize the discussion, *Car Life* is presenting a point-by-point comparison of the two cars, based on the road test of each.

First, a qualification: We selected the Corvette's most powerful engine (375 bhp, fuel injected) and brake (sintered linings, finned drums) options to give the Sting Ray an even chance. As can be seen by the respective data panels, the test weights figure out to 9.4 lb./bhp for the Cobra, 9.15 lb./bhp for the Corvette. Since the Cobra comes equipped with relatively large disc brakes, it takes the Corvette's full racing option brakes to match the AC's stopping ability. Another equalizing factor: The extra cost of the fuel injection engine, special brakes and knock-off wheels brought the Corvette's retail price to just about equal that of the Cobra.

CHASSIS—Corvette and Cobra share the same basic sort of structure, both depending upon the ladder type of frame layout with sub-structure to support the bodywork. In the case of the Sting Ray coupe, the sub-structure decidedly strengthens the torsional resistance of the chassis. The Cobra's twin-tube design dates back at least 12 years, but if not modern is at least lighter in weight. The Corvette frame, welded up of channel section rails, is overly bulky and heavy for competitive use although it provides a solid, firm platform from which the suspension works.

SUSPENSION—Here is one of the biggest differences between the two cars. They both have fully independent action at all four wheels but the Corvette is by far the more sophisticated in design. Where the front and rear wheel geometry of the Cobra tends to top wheels outward when the car leans during cornering, that of the Corvette tips them inward to keep the tires at a more perpendicular attitude. This, of course, improves the tires' ability to resist skidding, so, at least in theory, the Corvette can corner better than the Cobra. In practice, however, the Cobra has such stiff springs that the car leans hardly at all and cornering traction doesn't suffer too much.

Tires also enter into this picture. The Cobra has Goodyear G-8s as standard equipment, the Corvette Firestone DeLuxe Champions. The G-8s are an ultra-low profile, low cord angle tire developed from Goodyear's stock and sports car racing tires and give fantastic traction. By comparison, the Firestones seem just a bit wobbly and far less capable of resisting side-thrust. Whatever faults the Cobra has in its suspension design, the tires make up for in superior action.

BRAKES—The Cobra has 12 in. discs, the Corvette has 11.5 in. drums. Surprisingly, the Corvette's racing option brakes stopped just as well as, if not better, than did the Cobra's caliper discs. Also surprising, repeated stops from 100 mph produced fading in the Cobra's brakes (and some tail-twitching lock-up of the rear wheels), while the Corvette's brakes withstood the worst we could do without one whit of fade. The data panels show that although the Cobra would out-accelerate the Corvette to 100 mph by 0.8 sec., yet the heftier Corvette could decelerate from 100 mph in 3.1 sec. less time. The 0-100-0 test times for the two were: Cobra, 22.9 sec.; Corvette, 20.6 sec.

BODYWORK—The Cobra's body is of roller-formed aluminum panels, welded into larger panels and hand-finished into a light, smooth unit. The Corvette's body is built up of chopped glass fibers bonded by a polyester resin and formed in

CORVETTE vs. COBRA

male/female molds to maintain panel gauge and carbon-copy duplication. This fiberglass body is also light, but requires a stronger supporting structure, which adds weight. Although the fiberglass is weather and salt resistant, it is hard to finish smoothly and most Corvettes have a characteristic minute ripple on their finished surfaces.

INTERIORS—Traditional sports car interior design speaks out in the Cobra, where the instruments are simple round-faced dials set into a flat, leather-covered dash panel. All are easy to read at a glance (a big advantage when going fast) and are conveniently located in front of, and to the right of the driver. Ditto the Corvette, although there's a bit more "style" to the whole layout (which adds about 90 lb. extra weight, we're told). Seats on both cars are comfortable in that they fit the driver well enough to give him lateral as well as longitudinal support. The Corvette's seats seemed more softly sprung, and in the long run may be just a bit more comfortable.

DRIVE TRAIN—Both Cobra and Corvette have lightweight, durable, short-stroke V-8s for power and in both cars these engines develop considerably more horsepower than their passenger-sedan brethren. The Corvette, at 1.15 bhp per cu. in is perhaps the more finely tuned (the Cobra has 0.94 bhp/cu. in.), but the fuel injection system is far fussier than the single 4-barrel used on the Cobra. A better comparison here might have been the Corvette's 365-bhp 327 which has a single 4-barrel carburetor atop the fuel-injected's heavy duty block.

In both of these particular powerplants the crankshafts, pistons, rods and bearings are of more durable design and manufacture, the cylinder heads are opened up for better breathing, and the timing and carburetion adjusted to high-speed operation. Both have durability beyond ken, having competed and completed such well-known endurance tests as the Sebring 12-hour race.

The Cobra has the Ford-gearset variation of the Warner Gear T-10 transmission, the Corvette the Muncie 2.20:1 low 4-speed. These are virtually identical units although the Chevrolet gearbox is actually a redesign of the T-10 and probably a little tougher. The Corvette's gears, too, were better spaced, with the 1.28:1 third gear providing a particularly versatile about-town and passing gear.

Final gearing on the Cobra was 3.77:1, for the Corvette it was 4.11:1. However, because the Cobra had those ultra-low profile tires and the Corvette more normal size 15-inchers, the actual miles-per-thousand rpm figures came out comparable. The Corvette has a slightly higher piston-travel-per-mile figure, because of its longer stroke.

PERFORMANCE—Because of the similarities in final gearing, it was no surprise to the test crew that the performance figures for the Cobra and the Corvette were not very far apart. And, using realistic red-line figures for the engine, top speeds are almost identical. Both cars might rev past 7000, but to do so is to risk expensive breakage and the gain in acceleration is negligible, as it puts the next gear up too far past the torque peak.

On the road, there is really no clear-cut choice between the cars. It all comes down to how you like your performance: Firm, nimble and noisy (Cobra) or soft, supple and quiet (Corvette). On high-speed handling, the Corvette has a definite edge, being the more sophisticated in suspension design. But, at low speeds, the lighter Cobra can dart in and out of places that leave the bulkier Corvette panting and puffing, so the fun factor favors the Cobra.

In the final argument, then, the Cobra is just more pure fun to drive, although the Corvette accomplishes the same thing with less fuss and more finesse. ∎

If you'd just spent fifty miles twisting down a mountainside in this '65 CORVETTE, you'd be taking the words right out of our mouth about its new 4-wheel disc brakes.

Praise comes easily to Corvette's big caliper-type disc brakes. They're fade-resistant, heat-resistant, water-resistant and fuss-resistant.

Adjustment and maintenance? What adjustment and maintenance? These discs can get along swell without much help.

All this peace of mind is standard equipment on all '65 Corvettes. But even this doesn't complete the improvements.

We've added a new 327-cubic-inch V8. This new one is the best behaved 350-hp V8 you ever met. Loaded with all kinds of good things a high-performance engine ought to have. But tamed to a civilized purr. You can order it or a 300-, 365- or Ramjet fuel-injected 375-hp V8 on any Corvette.

As for appearance, it's brushed up. Smooth new hood, magnesium-style wheel covers, new grille design and functional front fender louvers. And a revised interior that treats you even more elaborately than before.

We didn't change all those lovely items you can specify, like 4-Speed fully synchronized transmission, cast aluminum wheels, Positraction—the lot.

All in all, Corvette for '65 adds up to a lot more car in every way. Which we think is the best possible reason for changing it.

Chevrolet Division of General Motors, Detroit, Michigan

CORVETTE ENGINEERING FOR 1965

by Jim Wright *Technical Editor*

WHAT CORVETTE LACKS in engineering *quantity* for 1965 it more than makes up in *quality*. The basic car is pretty much unchanged from the last two years. What changes have been made are in the areas of engine, driver convenience, and most important of all, braking.

For 1964, Corvette listed a choice of four engines — there are five for '65. All are based on the 327-cubic-inch block. Three are direct carryovers from this year: the 300-, 365-, and 375-hp versions. The basic 250-hp engine gets the 300-hp version's heads with bigger intake valves and ports, but with no increase in power. The brand-new engine is basically 1964's #3 mill — the 365-hp version — but modified slightly with a milder camshaft and hydraulic lifters. It's rated at 350 hp and gives the buyer an extremely potent street engine without the noise and rough idle of the 365- and 375-hp models. Cam timing places a large portion of the usable horsepower and torque curves down in the low and middle ranges, which also enhances the engine's street-ability.

The already comfortable and highly efficient cockpit becomes more so with the addition of an optional, telescoping, adjustable steering column. In '64 models, you could adjust the column with a wrench, but the new one has a collar set below the wheel that's easily manipulated by the driver. The three-inch adjustment coincides nicely with the available seat travel to let any driver find a comfortable driving position. Whether you prefer the arms-straight-out Italian style or the old use-the-wheel-as-a-chinning-bar style, both are easily obtainable. And if your taste lies somewhere in between, the extra-cost option will let you find it. This is an important step toward completely adjustable driving controls — one we hope other companies pick up.

Getting around to the brakes, we see that Corvette joins the likes of Jaguar, Mercedes, and a few select others in offering caliper-disc brakes at all four wheels. Development work on these brakes goes all the way back to 1955. In the beginning, Chevrolet did some joint work with Girling of England along these lines, only to find that at the time there wasn't a disc brake being manufactured that could cope with the Corvette's 3000+ pounds. Later development work (done in conjunction with Kelsey-Hayes) decided Chevrolet to handle the project all by themselves. Once they found all the answers, they handed the brake project over to GM's Delco-Moraine Division for manufacture.

With these brakes at all four wheels (instead of just at the front as is the usual practice), the '65 Corvette offers what'll come to be recognized as the ultimate in braking ability. We've driven this brake back at the proving grounds, and we've also spent some time studying various performance graphs made up during the pre-manufacture testing period. They're really sensational!

During performance testing, the brakes were subjected to just about every brake test imaginable — and then some. One of the engineers' favorite tests was a series of 20 .70%-G (23 feet/second/second) stops from 100 mph. All stops were made consecutively, with no time between them for cool-down, and each stop had to be made with absolutely no wheel skid. Existing drum brakes are lucky to survive one stop like this, but the Corvette discs were able to make all 20 — time and again — right to spec.

The discs' resistance to fade is pointed out by the fact that normal hydraulic-line pressure in this system (at 100 pounds' pedal pressure) is 500 psi. After the 20-stop series, line pressure had risen only to 625 psi. This is a very nominal increase. It can honestly be said that this system *is* fade-proof.

During other tests, the brakes have pulled test Corvettes down to swerve-free, no-wheel-skid stops at a deceleration rate of one G — 32.2 feet/second/second — a feat that many have believed physically impossible. What's more, the tests were performed on standard tires — not super-sized, gum-soft racing skins.

The Corvette we drove wasn't equipped with power assist, and it didn't need it. Required pedal pressures are less than those required on the majority of drum-braked passenger sedans. With 16 separate wheel cylinders — four to each caliper — the new system has a hydraulic ratio of 43.25 to 1. This combines with a pedal ratio of 4.45 to 1 for an overall ratio of 196.4 to 1.

Wear tests were made on a number of Corvettes that had made several cross-country trips to subject them to the stresses of everyday driving. After 40,000 miles, the cars were torn down. Based on the wear rate that resulted, the linings were found to have a projected wear rate of 57,000 miles for front linings and 127,000 miles for the rear. There was virtually no disc wear. When the linings do need changing, it takes only five minutes per wheel.

Except for the novel parking-brake arrangement (clearly shown in the photographs below), Chevrolet's system is very similar to existing ones. Each cast-iron caliper contains four hydraulic cylinders, which act on two sets of organic linings. Effective lining area is slightly more than 20 square inches per wheel: total 83.4 square inches. The cast-iron discs are 11.75 inches in diameter and 1.25 inches thick. These two solid braking surfaces are separated by cast-in radial fins, which make for rapid cooling during hard use. Total swept disc area — those surfaces rubbed by the linings — is 461.2 square inches. The linings are self-adjusting and are set up with zero clearance to the discs. This cuts down unnecessary pedal travel and also provides a constant cleaning action to the discs.

It may take another couple of years to get the costs down on this brake so it can be used on the standard Chevrolet passenger car. But after looking at performance comparison charts between the new discs and the old drums, they'd better not wait that long. /MT

Rear brake rotor or disc contains small brake drum in hat section. Arrangement is logical, easy solution for parking brake.

Rear dust shield also acts as backing plate for 6½-inch-diameter parking brake. Cables actuate this mechanical system.

Bottom view of cast-iron caliper shows approximate shape of linings. Groove in middle provides wear gauge, collects dirt.

Top view of caliper reveals pin that holds the lining in place. Changing the lining shouldn't take much more than five minutes.

PHOTOS BY BOB D'OLIVO, GENERAL MOTORS

Front rotors are same size as rears: 11-¾-inch diameter, ¼-inch thick. Internal radial fins help dissipate heat, promote cooling.

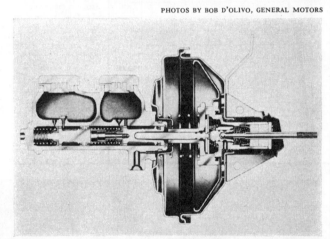

Non-power brakes need less pedal pressure than many drum systems. Optional power isn't really needed, yet it is available.

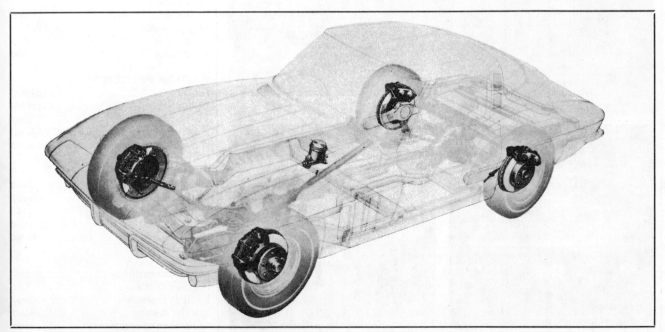

PHANTOM DRAWING SHOWS NEW DISC BRAKE LAYOUT. DEVELOPMENT OF THIS SYSTEM GOES BACK TO '55, INCLUDES MANY DIFFERENT DESIGNS.

CHEVROLET CORVETTE

Manufacturer: Chevrolet Motor Division,
General Motors Corporation
Detroit 2, Michigan
Price as tested: 5276.30 FOB, St. Louis

ACCELERATION

Zero to	Seconds
30 mph	2.1
40 mph	3.0
50 mph	4.3
60 mph	6.2
70 mph	8.3
80 mph	11.1
90 mph	13.7
100 mph	16.9
Standing ¼-mile	94 mph in 14.9

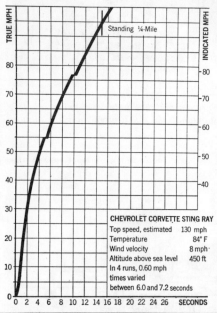

CHEVROLET CORVETTE STING RAY
Top speed, estimated 130 mph
Temperature 84°F
Wind velocity 8 mph
Altitude above sea level 450 ft
In 4 runs, 0.60 mph times varied between 6.0 and 7.2 seconds

ENGINE

Water-cooled V-8, cast iron block, 5 main bearings
Bore x stroke........4.00 x 3.25 in, 102 x 83 mm
Displacement.....................327 cu in, 5359 cc
Compression ratio.........................11.0-to-one
Carburetion..................Single four-throat Holley
Valve gear........Pushrod-operated overhead valves (hydraulic lifters)
Power (SAE)...........................350 bhp @ 5500 rpm
Torque..................................360 lbs-ft @ 3600 rpm
Specific power output......1.07 bhp per cu in, 65.3 bhp per liter
Usable range of engine speeds..600–5600 rpm
Electrical system....12-volt, 61 amp-hr battery
Fuel recommended.............................Premium
Mileage..10–18 mpg
Range on 20-gallon tank........200–360 miles

DRIVE TRAIN

Clutch..................10-inch single dry plate
Transmission.....4-speed all-synchro (Muncie)

Gear	Ratio	Over-all	mph/1000 rpm	Max mph
Rev	2.27	7.52	—10.5	—59
1st	2.20	7.40	10.7	60
2nd	1.64	5.51	14.4	77
3rd	1.28	4.30	18.4	102
4th	1.00	3.36	23.5	130

Final drive ratio.........................3.36 to one

CHASSIS

Channel-section steel frame, fiberglass body.
Wheelbase...98 in
Track................................F 56.8 R 57.6 in
Length..175 in
Width..69.6 in
Height...49.8 in
Ground clearance..................................7.0 in
Dry weight......................................2975 lbs
Curb weight....................................3180 lbs
Test weight....................................3500 lbs
Weight distribution front/rear............47/53%
Pounds per bhp (test weight)................10.0
Suspension F Ind., unequal-length wishbones and coil springs, stabilizer bar.
R Ind., radius arms and lower transverse rods, half-shafts acting as upper locating members, transverse leaf spring.
Brakes..........11.75-in discs front and rear, 461.2 sq in swept area
Steering.............................Recirculating ball
Turns, lock to lock..................................3
Turning circle......................................36 ft
Tires...7.75–15
Revs per mile.......................................760

CHECK LIST

ENGINE
Starting..Good
Response....................................Excellent
Noise...Good
Vibration..Good

DRIVE TRAIN
Clutch action.................................Good
Transmission linkage..................Excellent
Synchromesh action...................Excellent
Power-to-ground transmission..Excellent

BRAKES
Response....................................Excellent
Pedal pressure...........................Excellent
Fade resistance..........................Excellent
Smoothness...............................Excellent
Directional stability....................Excellent

STEERING
Response...Good
Accuracy...Good
Feedback...Good
Road feel...Good

SUSPENSION
Harshness control...........................Good
Roll stiffness...................................Good
Tracking..Good
Pitch control...................................Good
Shock damping...............................Good

CONTROLS
Location..Good
Relationship....................................Good
Small controls.................................Good

INTERIOR
Visibility...Fair
Instrumentation........................Excellent
Lighting...Good
Entry/exit..Fair
Front seating comfort.....................Good
Front seating room.........................Good
Rear seating comfort.........................—
Rear seating room.............................—
Storage space..................................Fair
Wind noise.......................................Fair
Road noise......................................Good

WEATHER PROTECTION
Heater..Excellent
Defroster...................................Excellent
Ventilation......................................Good
Weather sealing..............................Good
Windshield wiper action................Good

QUALITY CONTROL
Materials, exterior..........................Good
Materials, interior...........................Good
Exterior finish.................................Good
Interior finish..................................Good
Hardware and trim.........................Good

GENERAL
Service accessibility........................Fair
Luggage space................................Fair
Bumper protection..........................Good
Exterior lighting..............................Good
Resistance to crosswinds...............Good

CORVETTE STING RAY

You aren't suggesting that it's one of the best GT cars in the world. Or are you?

We know Americans can't build the best possible Grand Touring car. After all, that takes brilliant engineering and old-world craftsmanship of the kind only Europe can provide. It's heritage and breeding and all that sort of thing, old boy. Detroit is perfectly capable of producing blustering, big-engined buses that might pass as high-speed touring vehicles if they had proper brakes and handling. Yes, that's where they fall down. Squishy suspensions and incendiary brakes certainly make American machines intolerable for the connoisseur.

There is the Corvette, but it's hardly in the running when you consider absolute pinnacles of accomplishment in the field, such as the Aston Martin DB-5 or the Ferrari 250/GT.

It bears superficial resemblance to the authentic thing; the bucket seats are remarkably well-formed and comfortable, and the instrumentation is what you might expect on a GT car—with a large, clearly-numbered speedometer and tachometer placed right in front of the driver. But after all, there are dozens of cars with that kind of equipment that miss being GT cars by a mile.

Size is the standard that immediately comes to mind. The Corvette is simply too big.

Wait a minute. You say the Corvette is only one inch longer than the Ferrari and five inches *shorter* than the Aston Martin? But it has that enormous wheelbase . . .

Only 98 inches? The same as the Aston and only 3.5 inches longer than the Ferrari?

But wheelbase and overall length are only part of it. The Corvette is heavy like a truck.

Its curb weight is only 3180 lbs? The Aston Martin weighs 3450 in the same trim? But Ferrari claims only 2540 lbs for the 250/GT. Bulk and size are silly parameters in such a discussion anyway. The real question is engineering sophistication. That is the heart of any automobile of this sort. After all, the Ferrari and the Aston Martin cost over $12,000, and it's ridiculous to expect that a car less than half as expensive could compete. Take the suspension, for example.

The Corvette has a fully independent suspension system fore and aft? Yes, we know. The Ferrari and the Aston use *live* rear axles? Now what can that all mean?

Then there is the gearbox. The Aston Martin features an optional ZF five-speed unit that is unparalleled, and then, of course, the Ferrari's, though only four-speed, has a tremendous reputation. Muncie? You don't mean Muncie, Indiana? That may be where they make the

new General Motors four-speed, but who ever heard of a true Grand Touring car with a gearbox made in Muncie, Indiana? It's utterly barbaric. Maybe it does have one of the lightest, most positive linkages ever designed and near-perfect ratios, but it's nonetheless downright silly for *any* transmission to be built in Muncie, Indiana.

When it comes to powerplants, the question is academic. Corvette engines simply can't compete with the modern, overhead camshaft units of the Ferrari and the Aston Martin. Yes, yes, we know all about the Ferrari V-12 dating back to the 1940s, but it *is* one of the greatest designs of all time. And the Aston Martin engine is a lovely double-overhead-cam straight-six. Of course we know the big 365-hp and 375-hp Corvettes have a considerable edge in sheer power, but they're harsh, solid-lifter, semi-racing engines that hardly fit the mold of a smooth, silent GT powerplant.

A new engine for the Corvette? With hydraulic lifters? Three hundred-and-fifty horsepower at 5800 rpm? Silky-smooth? No rough idle? No pushrod clatter? One hundred more horsepower than the 250/GT? Sixty-eight more than the Aston-Martin? That all may be true, but consider the vast advantage the Corvette has in cubic inches. Let's talk in terms of engine efficiency. The Ferrari produces 1.36 horsepower per cubic inch, the Aston Martin gives 1.16 hp per cubic inch and the Corvette 1.072 horsepower per cubic inch. See what we told you? And let's not hear any nonsense about reliability and lack of temperament. When we're discussing *pure* machinery, mundane things of that nature have no bearing whatsoever. Speed in excess of 125 mph is essential for a car of this sort and on that count we'll have to give the Corvette a passing grade. With a 3.31 rear axle ratio, it will easily exceed that mark and, equipped with the optional 3.01 ratio, we grudgingly admit that the machine will top 150 mph without effort.

Anything that goes that fast must certainly be able to stop properly and we all know about those Detroit brakes, don't we?

Disc brakes? On the Corvette? On all four wheels?

As a matter of fact, we do recall reading some press release about Corvette discs, but surely they can't compare . . .

Vented discs, 11¾ inches in diameter? Great resistance to fade? An absolute revelation when compared to the old Corvette brakes?

That may be, but comparing old and new Corvettes is one thing; comparing the new Corvette with the Ferrari and the Aston is another. The Ferrari, for example, has an enormous 573 square inches of swept braking area and the Aston Martin has 468 square inches . . . which is by no means a meager quantity. By contrast the old Corvettes had a paltry 321 square inches.

The new Corvette disc brakes have 461.2 square inches of swept lining? You have the effrontery to suggest that they form one of the really outstanding braking systems available on a production car today? The entire question isn't worth arguing about. The Ferrari and the Aston Martin and other similarly pedigreed European cars are simply not in a class with the Corvette. They have what you might call "breeding."

Stop all that nonsense about the Corvette being as fast and as silent, as stable and as much in keeping with the grand touring concept as the other two. We don't want to hear how it might be argued that the Corvette is equally sophisticated from an engineering standpoint or that it might even be as well made. More reliable than an Aston or a Ferrari? Is nothing sacred?

There is more to an automobile than dull, simple economic value or its performance capability. There's tradition and there's . . . tradition . . . and then there's . . .

Anyway, you know what we're trying to say. **c/D**

PHOTOGRAPHY: TOM BURNSIDE

CORVETTE STING RAY Mk. IV

Now it's a fiberglass porcupine!

BY JERRY TITUS

WHILE GENERAL MOTORS CONTINUES TO PURSUE ITS NON-RACING POLICY and to promote the theme of Proving Grounds Development as the ONLY answer, it is about to put an engine in production that was developed specifically for racing. We are referring to the "Daytona" née "staggered-valve" née "Porcupine" V-8. It is an exceptional powerplant — about the latest word in pushrod design. It has more potential output-per-cube than anything currently rolling out of U.S. foundries. It will be available as an RPO in all the 119-inch-wheelbase passenger cars, and combined into a high-performance version of the Sting Ray.

Prior to our test of a prototype Mk IV Sting Ray, we had the opportunity to interview Zora Duntov, the man to whom Corvette development is entrusted. For the many who know this energetic man, we are pleased to report that while his hair is a little whiter, the twinkle in his eye is just as bright as ever. We pursued a line of questioning that made it tough for him to answer in a straightforward manner, as it concerned competition application of this high-performance vehicle. The use of the engine is pretty much evolutionary, but Chevrolet has at last taken pains to provide their customers who insist on racing with at least a chance of being competitive. The car has been submitted to the SCCA and the FIA with a list of options that are normally available to owners of other competition marques. This list includes oil cooler, exhaust headers, wide-rim wheels, big gas tank, and other goodies. Zora was particularly enthusiastic about the new Firestone tire that puts an exceptional amount

Below, Zora Duntov, in charge of Corvette development, explains points of the latest model. At right, the new "Ray" handles the wet banking of the Tech Center track without sliding.

of biting rubber on the ground without excessive diameter. It will, of course, necessitate fender flares to provide clearance, and it looks like the latter will be available in a do-it-yourself kit.

The car we tested had the 396-inch engine installed in a street version. Side-mounted exhausts were the only other optional hardware not normally available on the standard '65 models. This system uses the Budd in-pipe mufflers and makes enough noise to cause Chevrolet to list it as an "off-highway" setup. In other words, they won't be responsible for any tickets you get with it. With ice and snow in abundance, our test took place on the little "dog-bone" track at the Tech Center in Warren, Mich., rather than the Milford Proving

continued on page 119

1—The "396" fits the engine compartment with no difficulty, will do the same in earlier Sting Rays. Secondary portion of the large Holley carburetor has a high-speed system only.
2—Reshaped hood bubble—which also acts as air intake to carburetor—and optional side-mounted exhausts—with cast-aluminum guards—quickly identify latest, hottest of Corvettes.
3—Staggered valve arrangement in the interest of very direct porting gave cause for nickname to the engine: "Porcupine." Cast exhaust headers are efficient, but competition ones will be offered along with other performance options.
4—Cutaway reveals how direct porting is accomplished. Main bearings are four-bolted. Water-jacketing around plugs and valves is ample. As first version of this engine was 427 inches, it seems logical that this one could be resized to that amount with very little trouble.

Engine Bore & Stroke	Gross HP & Torque	Equipment	Comp. Ratio	Transmission	Rear Axle Ratio		Available as Positraction Axle Only.	
					Gen. Purpose Standard	Spl. Purpose or Mountain	Perform. Cruise	High Perform.
396 Cu. In. V-8 RPO L-78 4.094 x 3.76	425@ 6400 415@ 4000	HC Large 4-Bbl. Carb. Specl. Cam. Mech. Lifters	11.0:1	4-speed (2.20:1 Low)	3.70:1 (RPO G81)	4.11:1 (RPO G81)	3.08:1 3.36:1 3.55:1 (RPO G81)	4.56:1 (RPO G81)

3

PHOTOS: BOB D'OLIVO GM PHOTOGRAPHIC

4

63

America's only true sports car performs even more precisely with discs all around

by Bob McVay
Assistant Technical Editor

UNIQUE ON THE DETROIT scene—that's still Chevrolet's Corvette Sting Ray. It's as yet the only true American sports car. It makes no concessions for carrying more than two people and a reasonable amount of luggage, and it doesn't claim to be anything except what it is. ☐ Well, what is the Corvette? It's one of the hottest performing, best handling, most comfortable sports cars on the market, and some think it's one of the best looking as well. Each year since its 1953 introduction, the 'Vette has been getting more refined, faster, better handling, and gentler riding. ☐ This evolutionary change has made the Corvette easier to live with, but it didn't help out its stopping problem. The drum-brake setup just didn't keep pace with the car's skyrocketing performance — until 1965, that is. Disc brakes are big news on the automotive scene this year, but Chevrolet's Corvette builders don't believe in doing things halfway. ☐ All four wheels sport big, 11.75-inch-diameter disc brakes. After more than 1500 miles behind the wheel (from slow city traffic to 124-mph speeds on the race track), we've formed a pretty good opinion of these stoppers. They're just great — the final component that gives an already good sports car stopping power to match its go power. ☐ The new caliper discs use ventilated ribs. Four hydraulic pistons at each wheel press the organic linings against the discs and maintain a slight constant contact that keeps the disc wiped clean when driving in rain or under extremely dusty conditions. The

Corvette Sting Ray Road Test

amount of braking effort at the wheels is in direct proportion to the pedal effort. With four pistons per wheel (for a total of 16), plus a pedal ratio of 4.54 to 1, the overall ratio is 196.4 to 1 or about that of a conventional duo-servo drum-brake setup. But the new Corvette disc system has a much better feel and gives the driver a closer relationship to his car. Wheel lock-up is much easier to predict and avoid with this system.

Inside the rear disc brakes, Chevrolet has incorporated small, 6.5-inch-diameter drum brakes with 1.75-inch-wide conventional brake shoes. These serve as the parking brake.

The splash shields, the finned discs themselves, and even the wheel and wheel covers are designed to help cool the brakes. Judging from the system's performance, they do an excellent job. Our test car came with the power (vacuum) option. This includes a dual-circuit master cylinder, with separate front and rear hydraulic circuits.

In addition to our regular brake-testing procedures, we subjected the Corvette to five consecutive panic stops from 60 mph—without giving the brakes any chance to cool. All stops were swerve-free, fade-free, and the wheels didn't lock up. Our shortest stop was 137 feet and the longest was 166 feet, for a 153.65-foot average stopping distance for our five stops. It felt as if the brakes could continue to perform this way all day long.

We also used the brakes hard on long, fast, winding, downhill mountain roads, in city traffic, and during one of Los Angeles' infamous cloudbursts. After 1500 miles, in our opinion, the 1965 Corvette has the finest, smoothest-acting, and strongest set of stoppers available on any American automobile.

Important though they are, brakes are only one component of many that go into building a sports car. Suspension is also very important, and here's where the Corvette shines again with a refined, fully independent system. They've shifted more weight to the rear for better traction this year. Distribution is approximately 47/53. A five-crossmember frame mounts the 31-piece convertible body at four points on rubber biscuits, while the coupe body attaches to the frame at six points.

Independent, variable-rate coils up front (with a .687-inch-diameter anti-roll bar) are standard. A .94-inch-diameter bar comes with the stiffer suspension option. A nine-element, transverse-leaf, variable-rate spring is mounted at the rear, with strut rods and torque control arms. Direct double-acting shocks, with special freon-filled bags inside, are used at each corner.

This car was definitely built with fast, safe driving in mind. It should prove, in the hands of an experienced driver, to be one of the quickest in captivity. At low speeds around town or on rough roads, ride is a bit choppy, but the Corvette is far more comfortable and easy to live with than many other sports cars. As speed increases, so does riding comfort. For high-speed travel over any kind of road, it'll be hard to find a more comfortable sports car.

The Sting Ray feels almost perfectly neutral in corners until pushed right to the limit. At that point, there's some understeer and subsequent carb flooding. But the Corvette can be cornered fast and flat on the track or on the road. Once set into a bend, we found that only slight changes in the throttle setting could alter our course, without our moving the wheel. But at all times, we had a feeling of complete control. The Corvette could be drifted, skidded, or just driven very fast through any variety of turns with a high degree of safety and control. It's a car that a good driver can really fall in love with, and one that a bad driver, or an over-enthusiastic

Roadster shows versatility, with three distinctive body styles when optional hard top is ordered. Definitely not a one-man job, taking off top can be a chore. Once in place, it rattled and required lots of muscle to lift up in order to get into luggage compartment.

Fully independent suspension provided excellent traction on the road, off the road, and on the track. Front tread is increased by .5 inch; rear by .6 inch for 1965. Car showed minimum of lean and understeer and maximum control at speed.

PHOTOS BY PAT BROLLIER

CORVETTE continued

Corvettes use a Delcotron, diode-rectified, air-cooled generator and a 12-volt battery. Engine and components are easy to reach for routine service, but battery is well hidden.

Husky 11.75-inch-diameter by 1.25-inch-thick ventilated discs give Corvettes fantastic stopping power. Without a doubt, they're the best set of stoppers on any American car.

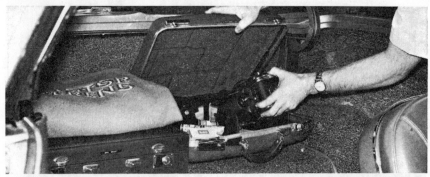
Corvette's coupe claims 10.6 cubic feet of luggage space, but the roadster's compartment is a bit smaller. When the top is down and hard top installed, luggage is difficult to reach.

Proving its mettle on California's winding mountain roads, our Corvette proved itself a fast, controllable sports car that was comfortable, quiet, and a joy to drive and live with.

one, can get himself into trouble with just as quickly.

Positraction was a big help in preventing wheelspin in corners, during acceleration, and especially on wet pavement. With power steering and brakes on our test car, control meant a very light touch. The rear felt just a bit twitchy on wet or dirt roads, but once we got the feel of the car, it proved able to do anything we asked.

The 300-hp or the standard 250-hp version of the Corvette's multi-horsepowered 327-cubic-inch V-8 won't make anyone a king at the local drag strip. But you can rest assured that if you aren't satisfied beating most of the people most of the time, you can always order a hotter version. The 350-hp "street" version and a 365-hp option top off the carburetor-mounted engines, while fuel injection injects a fire-breathing 375 hp into this fiberglass beauty. And that one ought to be fast enough for anyone. If it isn't, just wait for the 396-inch V-8 soon to be introduced.

Inside the Corvette, there's a very functional dashboard, with big, legible gauges. Everything's within easy reach, especially that beautifully smooth, precise, four-speed transmission and its handy, short chrome lever.

Six acrylic lacquer finishes and a combination of interior tones give buyers a wide color choice. Convertible tops come in black, white, and beige. And there's a coupe body, too. New seats for 1965 give more lateral support during hard cornering plus somewhat better support than last year's seats. New molded door panels boast integral arm rests.

Our "commuter" Corvette, with its 300-hp engine, was a very easy car to live with — much more so than former fuel-injection models we've driven. It started easily, was completely docile in traffic, needed only first and fourth for most normal demands, and it was very quiet. On the other hand, it gave excellent performance for a 3280-pound car when we used the 5500-rpm red line and the four-speed gearbox to the fullest extent.

Most normal driving gave between 11 and 14 mpg of premium gas, but steady highway cruising boosted this up to around 17 mpg with the standard 3.36 rear axle. Optional axles range from 3.08 to 4.56.

In standard form, Corvette's 1965 offerings are smooth, quiet, comfortable sports cars, capable of staggering performance depending on what engine you order. A better all-around sports car would be hard to find at any price. Here's a car a man can really enjoy driving and living with. We loved it. /MT

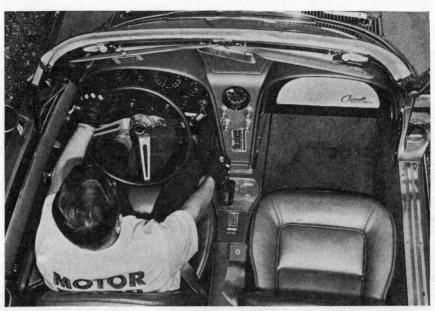

Full instrumentation in white-on-black dials is refreshing after mazes of warning lights.

Good layout makes this one of the most comfortable of all cockpits we've had the pleasure of testing. Relationship of controls is good. All are within easy driver reach.

FIRST YOU HAVE TO KNOW THE PROCEDURE, THEN TOP GOES UP OR DOWN WITH EASE. THE LOUVERS AND WHEEL COVERS ARE NEW THIS YEAR.

CORVETTE STING RAY
2-door, 2-passenger convertible

OPTIONS ON CAR TESTED: 300-hp V-8, 4-speed manual transmission, Positraction, power brakes-steering-windows, removable fiberglass hard top, AM-FM radio, power antenna, tinted glass, whitewalls.
BASE PRICE: $4106
PRICE AS TESTED: $5279.35 (plus tax and license)
ODOMETER READING AT START OF TEST: 2000 miles
RECOMMENDED ENGINE RED LINE: 5500 rpm

PERFORMANCE

ACCELERATION (2 aboard)
- 0-30 mph.....................................3.1 secs.
- 0-45 mph.....................................4.6
- 0-60 mph.....................................7.5

PASSING TIMES AND DISTANCES
- 40-60 mph..........................4.1 secs., 299.3 ft.
- 50-70 mph..........................4.3 secs., 378.4 ft.

Standing start ¼-mile 15.8 secs. and 90 mph
Speeds in gears @ 5500 rpm
- 1st .... 50 mph   3rd .... 90 mph
- 2nd .... 70 mph   4th .... 124 mph
                                   @ 5200 rpm (observed)

Speedometer Error on Test Car:
| Car's speedometer reading | 30 | 45 | 51 | 61 | 71 | 81 |
| Weston electric speedometer | 30 | 45 | 50 | 60 | 70 | 80 |

Observed miles per hour per 1000 rpm in top gear..............23 mph
Stopping Distances — from 30 mph, 30.5 ft.; from 60 mph, 137.0 ft.

SPECIFICATIONS FROM MANUFACTURER

Engine
Ohv V-8
Bore: 4.00 ins.
Stroke: 3.25 ins.
Displacement: 327 cu. ins.
Compression ratio: 10.5:1
Horsepower: 300 @ 5000 rpm
Horsepower per cubic inch: 0.92
Torque: 350 lbs.-ft. @ 4000 rpm
Carburetion: 1 4-bbl.
Ignition: 12-volt coil

Gearbox
4-speed manual, all-synchro, with reverse lock-out; floorshift

Driveshaft
1-piece, open tube

Differential
Hypoid, semi-floating (Positraction)
Standard ratio: 3.36:1

Suspension
Front: Independent, with coil springs, unequal-length upper and lower control arms, direct-acting tubular shocks, and anti-roll bar
Rear: Independent, with fixed differential, transverse, 9-leaf springs, lateral struts, and universally jointed axle shafts, radius arms, and direct-acting tubular shocks

Steering
Semi-reversible recirculating ball nut; linkage-type power assist
Turning diameter: 39.9 ft.
Turns lock to lock: 3.4 or 2.92 (optional)

Wheels and Tires
5-lug, steel, short-spoke spider wheels
7.75x15 4-ply tubeless rayon whitewall tires

Brakes
Hydraulic 4-wheel caliper discs with integral power assist and dual-circuit master cylinders
Front: 11.75-in. dia x 4 hydraulic pistons & 2 shoes per disc
Rear: 11.75-in. dia., with 6.5-in. dia. x 1.75 ins. wide drums for parking brake
Effective lining area: 83.4 sq. ins.
Swept disc area: 461.2 sq. ins.

Body and Frame
31-panel fiberglass body, attached to welded steel, ladder-type frame
Wheelbase: 98.0 ins.
Track: front, 56.8 ins.; rear, 57.6 ins.
Overall length: 175.1 ins.
Overall width: 69.6 ins.
Overall height: 49.8 ins.
Curb weight: 3280 lbs.

JAGUAR XKE
Most Overrated?

CORVETTE STING RAY
Just a Plastic Chevy?

For many years Jaguar offered the most performance for the dollar of any sports car in the world. Its competition record kept fans loyal and sales up. Then, in the United States, the Corvette V8 emerged to challenge Jaguar both on the track and on the street as a prestige personal automobile. A radical design change in 1961 brought Jag back strong in the glamour group although it hasn't been able to score in competition. So also has the Corvette been eclipsed in that department in spite of a similar re-design in 1962.

☐ Now we have two fast, sporting *Gran Turismo* coupes of approximately the same character which, comparably equipped, are within the same price bracket. They represent attainable luxury concepts for many people.

☐ Which car, Jaguar XKE or Corvette Sting Ray, deserves the most consideration from the buyer who seeks this type of transportation?

JAGUAR PRICES

■ One big factor which can complicate a selection is the number of equipment choices which are part and parcel of the Corvette. In the true Detroit manner there are stated options as long as a laundry list for a family of eight. On the other hand a Jaguar is a Jaguar is a Jaguar in the essentials and about all the purchaser can whittle off the nominal $6,000 ticket is $175 for the chrome wire wheels, $50 for tinted glass, $50 for white sidewalls, and $15 for seat belts.

☐ Let's first consider what you get for the money.

☐ In the main — although $6,000 is a good piece of change to spend on impulse buying — most cars of this type are sold with the load, but a stripped XKE Coupe can be had for $5,685 P.O.E. West Coast. The roadster lists at $5,585. Prices would be $60 less if you could prevent the distributor from attaching the 'optional' bumper overriders, but these seem to be affixed at the dock and you may as well live with them. Included, for these figures, are such items as 265 bhp engine, four speed fully synchronized gearbox, limited slip differential, center lock wire spoke wheels (painted), four wheel power-assisted disc brakes, rack-and-pinion steering, adjustable steering wheel, windshield washers, 3-speed wipers, wood-rimmed steering wheel, leather bucket seats, full instrument panel, padded sun visors, heater, courtesy and map lights.

☐ Incidentally, painted wire wheels, although stock, are a special order. All XKEs coming to the States have the chromed type and probably 95% are sold so equipped.

CORVETTE OPTIONS

■ The Sting Ray Coupe has a basic list price of $4,321.

• Jaguar front suspension is by forged control arms and torsion bars. Corvette uses stampings, coil springs.

• Twin overhead cam Six has been Jaguar's power for over fifteen years. Alternator is new in 1965 models.

The roadster is $4,106. For this sum the purchaser gets a 250 bhp engine, 3 speed transmission, heater, vinyl bucket seats, adjustable steering wheel, 4 wheel disc brakes, seat belts and courtesy lamps. This would also be a 'special order.' ROAD TEST shoppers could not find a dealer in the Los Angeles metropolitan area (including Harry Mann, the world's largest Corvette dealer) who could remember delivering such a car. However, for the $1,300 price differential between this and the basic XKE, practically all of the assorted extras which are found on the majority of Corvettes can be applied. Without many of them, the Corvette is not competitive except in road holding and braking — leaving aesthetics entirely out of the discussion.

☐ The first step up the Corvette ladder is an alternate transmission. Chevy's 3-speed isn't the worst box in the world — in fact it compares rather favorably with the transmission just abandoned by Jaguar, being just about as noisy and having a non-synchro first gear — but the gap between first gear ratio (2.58) and second (1.48) is rather wide for sporty driving and the necessity for coming to a complete stop before engaging low gear is a nuisance in a non-economy car. The choice is between a Powerglide (at $199) or the all-synchro four speed with 2.56, 1.91, 1.48 and 1.00 ratios, (for $188). Another four speed with close ratios, (2.20, 1.64, 1.28, 1.00) is available only with 365 and 375 bhp engines. Powerglide can be had only with the 250 and 300 bhp engines.

☐ Engine choice is next. Chevy's 327 cubic inch V8 with 4.00 inch bore and 3.25 inch stroke is the sole offering but it comes in four stages of tune, as indicated on the accompanying chart. The significant increase of 50 bhp from the standard engine to L-75 is not easily accounted for. Cam timing is the same and the only difference is in carburetion. A Carter WCFB 4 barrel with 1.43 inch venturis used on the standard engine and an aluminum Carter AFB with 1.56 inch primary and 1.68 inch secondary venturis comes with L-75.

☐ The next option, L-76, has a number of changes: Cam, pistons, rings, bearings, valves, compression ratio, carburetion and exhaust system to justify its power increase. The L-84 gets it rating from the use of Rochester Fuel injection. Ten extra horsepower hardly seem to rate a $409 price differential, but if you intend to go into competition, the injected engine is mandatory because it is not affected by attitude whereas engines with this type of carburetor have a tendency to starve or flood (or both) under racing conditions.

☐ Although disc brakes are standard on all Corvettes, power is extra. This $43 item is not necessarily for little old ladies. Jag has servo-assist and the 'purists' approve it. Pedal pressure is high with disc brakes and, whether you are racing or touring, heavy application can be tiring.

CORVETTE ENGINE SPECIFICATIONS

Engine	Comp. Ratio	Carburetor	Torque	Horsepower
Standard	10.5 - 1	WCFB	350 @ 2,800	250 @ 4,400
RPO L 75	10.5 - 1	AFB	360 @ 3,200	300 @ 5,000
RPO L 76	11.0 - 1	Holley	350 @ 4,000	365 @ 6,200
RPO L 84	11.0 - 1	Inject.	350 @ 4,600	375 @ 6,200

Likewise, Power steering is found on most Sting Rays because its front suspension makes it pretty much like a truck at low speed. The limited slip differential isn't included in the base price either, but it is necessary if you're specifying anything but the 250-Powerglide.

☐ By the time you add $81 for real leather upholstery on the bucket seats you're equipped the car in a manner equivalent to the basic XKE with the exception of center-lock wheels and premium tires. You can't get wire wheels from the Chevy factory, but a set of light alloy knock-off disc wheels lists for $322.80. Nylon tires are $16 more, and they compare reasonably with the Dunlops fitted to the XKE. (If you just want whitewalls, they're $32). Tinted glass, power windows, radio, air conditioning, transistorized ignition and such are filligree which don't really count in the analysis.

☐ So, here's the financial picture:

☐ *Jaguar XKE Coupe* with painted wire wheels, bumper guards and seat belts..$5,700

☐ *Sting Ray Coupe* with 4-speed, L-84 engine, power brakes and steering, limited slip, leather, nylon tires, knock-off wheels...$5,649

☐ If you want to drop back a notch on the engine option, the Corvette can list for $5,241. And if you want to extract all the prestige which goes with an XKE you'll have to take the chromed wheels and that means, $5,875.

☐ Is a $5,241 Sting Ray competitive with a $5,700 XKE in performance?

☐ Yes, but just.

☐ The cars are extremely closely matched in acceleration, (both attaining approximately 94 mph at the end of a standing quarter mile) and, unless pushed to the limit of their capabilities by talented drivers will run nose to tail over nearly any road. Finally, however, in top speed the Jag will show a slight superiority and it actually takes the injected Sting Ray to beat one. With the L-84 engine though, there is a noticeable difference and the Corvette emerges better on the basis of lots of horsepower applicable at all times.

☐ Two cars, however, could hardly be more evenly matched in appeal and purpose. What are the features which can lead the buyer to select one or the other?

DESIGN & EXECUTION

■ The biggest thing Corvette has going for it is packaged into: tremendous reliability; parts and service obtainable in any city, hamlet or village and domestic origin. Its biggest drawback is strictly one of aesthetics... not necessarily in external appearance, (which is a matter of taste and appeal ROAD TEST does not presume to pass on) but in the subtle lack of elegance in design and execution. This is a hard matter to define, but simply stated it is the invisible factor which causes night club parking lot attendants to put E Jags in the front row and stick Sting Rays in any convenient hole.

☐ This difference in design concept is apparent throughout both cars. It is the difference in suspension control arms which are forgings or stampings; rack and pinion or recirculating ball steering gear; polished cast aluminum cam covers or drawn steel; genuine wood rimmed steering wheel or an almost undetectable imitation thereof. Understand, this is not necessarily a brief for the "classic" method of accomplishing things, but merely to point out how the two cars can get the same job done in different ways ... with either great approval or virulent disapproval from

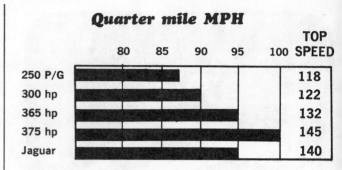

critics on both sides. The manufacturing attitude which spawns the Corvette is one which has made our automobile industry a giant: Design something; build it; sell it; when it goes wrong in service, change it to make it work; build the change into the next model.

☐ With the Jaguar it is: Design something as nearly right as possible; build it; if some one buys it who can't accommodate himself to the design, let him buy something else; if something goes wrong, it shouldn't have because the design was right.

☐ These two attitudes can explain why the S.U. Carburetor and fuel pump are still in production in England and why fuel injection is almost a stock item here.

☐ Jaguar's greatest attribute is that it is a sum of these very factors which are by-passed in America. Factors which make it, in its way, a superb car to own and drive: Handling, performance, execution of design. On the negative side are: gaps in the service chain; a reputation for touchiness in tuning and the well known fact that many minor parts have the lasting quality of cobwebs.

EXTERIOR & INTERIOR SIZE

■ In a side-by-side comparison the cars' differences and similarities are accentuated. They are within 1/16 inch of being identical in overall length, although the Corvette's wheelbase is two inches greater. It is also wider, by four inches and two inches higher. Body lines make it appear much blockier than the Jag and there is actually more interior room in the Sting Ray. The seat-to-roof dimension in the Corvette coupe is two inches greater than the Jag coupe. Another 1½ inches in headroom can be gained by picking the Corvette convertible-hardtop, but you'll lose an inch in the equivalent Jag. Effective headroom is improved in the Sting Ray by virtue of the fact that the seats are raked back 28°. The Jag's buckets are more nearly vertical and in the new 4.2 model, are more padded than before, bringing passengers closer to the instrument panel.

☐ Neither car has provisions for other than two passengers although small children, or an uncomplaining adult, can be carried in the nominal luggage space of the coupe. The Jag coupe comes off better here, more luggage can be stowed and is more easily accessible than the Sting Ray because of its rear deck door. The XKE roadster is less well blessed. Its capacity is strictly limited, whereas the Corvette roadster has ample room for a touring twosome's luggage.

CREATURE COMFORTS

■ Because of the intrusion of the transmission tunnel neither of these cars will win any awards for stretch out room. It is completely non-existent on the driver's side and

HOW THEY COMPARE

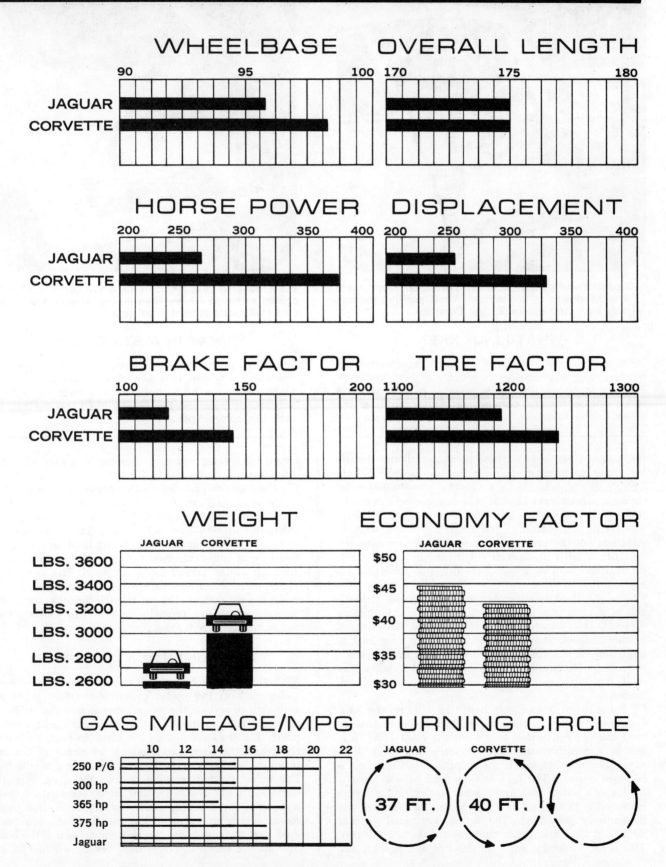

• Six-foot driver's head touches roof of Jaguar coupe, there is less headroom in convertible. Wheel position permits slightly more arm extension than Corvette.

WHAT IS AN XKE?

XKE is the designation of Jaguar's sports-touring model. It is available in two body styles: fastback coupe and convertible. A removable plastic hardtop is an option for the convertible. Other options are given in the text.

• Six foot driver has ample headroom in Corvette, right leg can be extended farther than in Jag. Tunnel makes both cars inadequate in stretch-out room.

WHAT IS A STING RAY?

Sting Ray is the designation of Chevrolet's Corvette sports-touring car. It is available in two body styles: fastback coupe and convertible. Either the folding soft top or a removable plastic hard top is standard with the convertible.

---

the necessity for assuming one position with the throttle leg is apt to become annoying to some people. ROAD TEST's Editor, who drove a Corvette some 4,000 miles through Mexico and Central America competing in the *Rally Centro America*, says that, "you either find this position to your liking or you forget it." Being short in stature is not the answer. As many 5'8" drivers objected as did their taller brothers. If either car had a hand throttle, it would help by allowing the driver to straighten his leg from time to time.

☐ For the short haul, bucket seats are equally comfortable in the two cars, although seating position is noticeably different. Over long distances, however, the more relaxed posture of the Corvette is better but there is some lack of thigh support for the tall driver. Jag seats can be made to recline a bit more by cutting the seat back bracket. The small flip-stop provided only permits an infinitesimal adjustment.

☐ If the driver is six feet in height he will find the XKE Coupe an extremely close fit and will discover that he can't use the convertible at all. On the other hand, 6'2" can be acceptably comfortable in the Sting Rays. Both adjustable steering wheels are helpful. The Corvette's is easier to operate. The Jag's lets the arms extend more.

☐ Jaguar's pedals are narrow and close together, although improved in this regard over past years, but are obviously thoughtfully positioned. The accelerator pedal is bare metal and tilts back, rather than forward, as in other cars. This is a comfortable angle for the foot and heeling-and-

toeing for braking and gear changing is a natural action. The Sting Ray is also well laid out.

☐ The Corvette's heating and ventilating system is typical of the American car; it works and is oversize. Kickpanel vents admit great quantities of fresh air and it is blower-exhausted through roof pillar vents. The XKE does not come off so well. Jaguar is still trying to fit an inadequate Smiths unit into a contrived system and there isn't enough fresh air at the critical lower leg level. The swing-open rear quarter windows create enough negative pressure to make a draft through the car but the tunnel radiates considerable heat. On a typical summer day in Southern California, the Corvette is more comfortable.

INSTRUMENTS & CONTROLS

■ Both cars have big, almost oversize, speedometers and tachometers as the center of interest. Other gages are scattered around on the same panel before the driver in the Sting Ray. In the XKE they are lined up in a center panel. The Jag's small dials are easier to read, (although the eyes must be moved farther), by virtue of the fact that the covering glass is flat and relatively non-reflective. For some incomprehensible reason, the glass over the Corvette's gages is concave and highly reflective, destroying readability in many situations. Jag also gains points for its switch arrangement, low fuel level and low brake fluid level warning lights. Another Jaguar plus is the headlight flasher which activates high beams, whether or not

• Luggage space is good in XKE coupe, access is via large rear deck opening. Barrier keeps small objects out of passenger compartment.

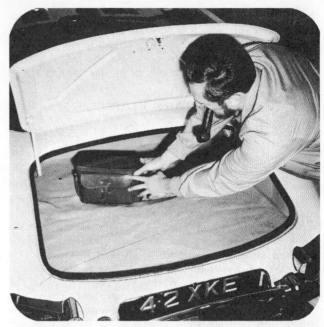

• Carrying capacity of Jag. convertible is minimal. Spare tire is under floor in both cars. Corvette's spare is easier to remove and replace.

the lights are on, enabling you to signal another driver of your intention to pass without honking the horn. The headlight-dipper is on the instrument panel. Some drivers prefer the conventional foot switch of the Corvette.

☐ Instrument location and readability is about all that Jag has going for it in this department however. The Smiths instruments, notably tach and speedo, are highly unrealiable and one can safely predict a short, happy life for them. The problem is in the tiny gears and components plus the fact that there is no oiling wick and the future owner is at the mercy of whatever young lady assembled the gage as to whether it was properly oiled in the first place. On the other hand, Corvette's Delco stuff has gears somewhat like those found in washing machines and they can be lubricated in service.

POWER TRAIN

■ Jaguar's new 4.2 engine (258.4 cubic inches) is a still bigger bore version of the same six cylinder double overhead cam powerplant which has been in service since 1949. Approximately 100 more horsepower is being wrung out of this engine than in its original form and, while it is not overstressed, it works pretty hard. (The bhp rating is fairly accurate, incidentally.) Oil consumption (500 miles per quart) is considerably greater than most American car owners expect and the average XKE owner has some sort of complaint about carburetion or ignition problems at some time. The S.U. carburetors are of a unique variable-venturi type which is more time-consuming to adjust and synchronize than the more conventional fixed-venturi type. Highly-dyed, gummy gasoline, lots of cold running on short trips and lack of attention causes them to give more trouble than, say, the Carter or Holley on the Corvette. Pinging, sooting and breaking down of plugs are common complaints, generally caused by too much use of the choke, excess spark advance or too lean a mixture favored by most mechanics. However, some blame can be placed on the advance curve in the distributor itself which does not permit enough retard at low engine speeds which drivers are prone to use simply because the engine pulls the car uncomplainingly.

☐ The XK engine has a good record for longevity but in pre-1965 models the Lucas starter and generator required replacement long before those in the cheapest American car. This year an alternator has supplanted the generator and a constant-mesh starter has finally done away with the antiquated Bendix drive which rammed gears together at each time the button was pressed. The electric fuel pump was submerged in the fuel tank in earlier models and gave no trouble in comparison with the S.U. found in so many British models. It has now been moved out into the open and remains to be proven in this situation.

☐ An important thing for the Jag owner is to find a good mechanic and take the car to him regularly. The mechanic won't be as easy to locate as a good Chevy man. The average Jag mechanic will be more intelligent and creative but he will also lack the simple virtue of trusting factory specifications implicitly, which is what is needed.

☐ Jaguar's new all-synchro gearbox is much smoother than earlier models, has a shorter throw to third (preventing knuckle-banging on the instrument panel) but is still full of whine. Jags have little transmission or rear end trouble but when problems occur, they're more expensive to repair than Chevrolets.

☐ The old reliable 327 V8 in the Corvette isn't as glamorous or as handsome as the XK, but it seldom gives any static and has good power . . . not as much as advertised, unfortunately. The 375 bhp comes closest to living up to its press notices, but the others are rated about 20% too high. The art of tuning the fuel injectors isn't as widely practiced as the less sophisticated carburetion systems, but procedure is spelled out in the shop manual available to every dealer and the unit seldom goes bad except from inexpert tinkering.

☐ Corvette's four speed gearbox is extremely smooth, quiet and rugged. Provision is made for an adjustment on the shift lever to make the throw shorter or longer, a nice touch, but either way it is accurate and easy. Powerglide, while strong and willing, is only a two-speed and loses some performance for this reason. Like the Jag, few troubles here, but when they come, parts and services are at hand.

THE CARS IN USE

■ There is quite a difference in the road feel of the two cars, although they are the same size. The Jag definitely feels smaller and more like an extension of the driver. A principal factor is rack and pinion steering, which, regardless of ratio, gives a more definite feel of the road than any other type. There is almost no play at the steering wheel in dead ahead position, so a slight movement makes something happen up front. The Jag's shorter turning circle also contributes. In handling, they are equally matched although one seems to work a little harder in pushing the Corvette to maximum. In brakes, the nod must go to Corvette because of reserve. The XKE has just enough brakes, no more. They're great, smooth and precise but on the edge. Corvette's discs surprise every new driver with controllability. They'll go as far as tires will permit.

☐ Both cars have a considerable amount of freeway hop, with the Corvette being worse than the Jag. (Some stretches of old concrete with tar strips every 20 feet will rattle your brains in a Corvette.) Ride otherwise is about even (speaking of the stock suspension in Corvette ... the optional HD suspension is for racing). Noise level is slightly higher in the Sting Ray coupe whose plastic body amplifies more than the Jag's metal shell. Again, wire spoke wheels seem to transmit less road noise than solid discs, so this may enter into it. The main drone sound from the XKE is exhaust resonance.

☐ Windshield wipers and washers on the Jag are far superior to the Corvette, but, like all the rest of its genuine rubber components, the blades don't last very long. Jag weatherstripping, moldings and so on soon disintegrate unless carefully maintained with a rubber dressing. The Sting Ray uses neoprene which is more impervious to weather. The small vent-return tube in the gas tank of the XKE succumbs within the first thousand miles and the car smells strongly of fuel until it is replaced with plastic. The chances are that the distributor drive dog at the back of the cam will shear pretty early, maybe before the instrument gives up. Turn signal switches are quite apt to stop operating in the first year and if the car is left in hot sun daily, the imitation leather will begin to peel off windshield posts and dash. If the owner operates at factory-recommended tire pressures and only motors gently, tires will wear in the center of the tread too soon. If he drives pretty hard, fine. If he reduces the pressure about 4 pounds ride will be good, wear will be good, but he can't play Dan Gurney on corners.

SUMMARY

☐ The top two Corvettes are a good match for the XKE in performance, comfort, convenience, handling, braking and safety. They are better suited to the person of above average height and require relatively less maintenance to remain in good operating condition. Parts and services are easier to find in most areas and more reasonable in cost than the Jag. As Jaguar's current ads say, "This isn't a car you can buy and forget..." It requires attention to details. But, with such attention, the automobile will present an elegant appearance long after the Corvette's fiberglass body has become a little frowzy.

☐ You can forget list price, except as a place to begin, on both cars. In times of plentiful supply, Jag dealers will give $400 to $500 off list on the $6,000 piece. Corvette dealers will do even better if you catch the right one at the right time. On resale, you're going to take quite a depreciation, period. At this point in time the year-old, non-injected model will fetch something over $3,000 wholesale. Jags have been about $200 to $100 back of book for nearly a year, so around $4,000 might be expected. Unlike Corvette people, Jaguar owners seem to keep their cars for longer than the two-year period which is the normal trade in period for many buyers because of basic satisfaction and lack of model change. This helps the depreciation factor since the curve flattens out with time. If you are the meticulous type you can have a cream puff Jag several years from now and recover pretty well. If you are a person who buys a car to drive and not to play with, better take the Corvette. There isn't anything on the market to compare with the fairly-stripped, 350 hp, 4-speed Corvette which lists at $4,660 and it must be considered a best buy in this category. However if you want the good things you'll have to escalate to about $5,500 on a deal and then you get into the coin-flip area.

☐ ROAD TEST rates the FI Corvette and XKE as equal values and feels that the determining factor is strictly one of buyer personality.

• **Jaguar instruments stretched across panel are more readable than Corvette's under-hood grouping.**

You've heard the rumors.

Now hear this...

There is a Turbo-Jet 396 from Chevrolet.

You've heard it's first cousin to the mysterious Mark II Chevrolet engine? It is. You've heard how well it breathes and how strong it runs? It does.

We call the beast Turbo-Jet 396, after its displacement. There's a high-output version you can order in Corvette. Bore: 4.094. Stroke: 3.76. Compression ratio: 11:1. One 4-bbl. Mechanical lifters. Bhp: 425. Torque: 415 lb.-ft. That's what. Now, why.

Breathes deep, free—Combustion chambers in the head are modified wedges with a large quench area for cooling the charge and a close-to-center spark plug for better combustion control. Independent ball-stud rocker arms allow tilting the intake and exhaust valves both lengthwise and crosswise to the engine's axis.

Tipping the inlet valve toward its port permits less restricted induction along a fairly uniform cross-section. Similarly tipping the exhaust valve allows a gentler outlet radius and less restricted exhaust gas flow. Finally, tilting inlet and exhaust valves *away* from each other on *two* axes unshrouds them for maximum volumetric efficiency.

What else is new—The short, rigid block for our Turbo-Jet 396 is strengthened above each bearing support by thick bulkheads. There are four bolts for each bearing cap instead of two. Main and connecting rod journals are specially hardened.

Pistons for this 425-hp version are aluminum impact extrusions developed by Chevrolet for high-output engines. Thermal expansion is controlled by a barrel-contoured skirt, eliminating those long slots below the oil-control ring and strengthening the piston. Connecting rods are beefed up, too.

Mostly, though, it's the breathing that makes our Turbo-Jet 396 big news. Deep on intake. Free on exhaust. You know how shrouding can strangle an engine.

Not this one.

Chevrolet Division of General Motors, Detroit, Michigan

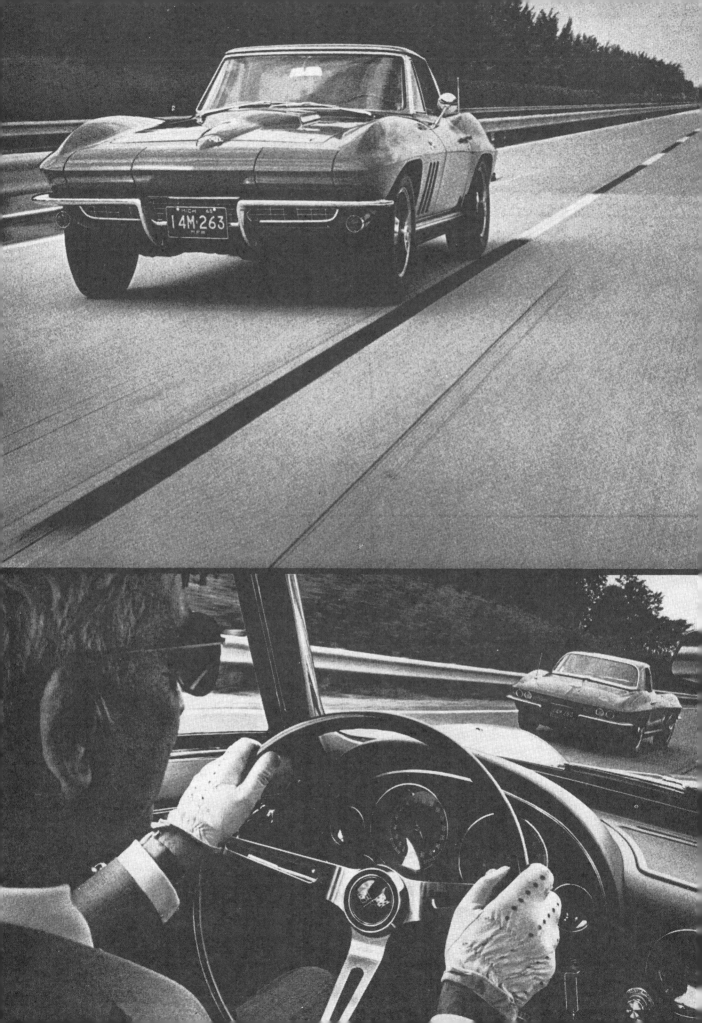

CORVETTE STING RAY 427

Next best thing to a psychiatrist—with electric windows, AM-FM, and about a jillion bhp

Son of a Gun—just what the Corvette needed, more power! Chevrolet has replaced last year's top-of-the-line 396-cu. in. engine with a 427. According to the factory, horsepower remains unchanged at 425, but torque has gone up from 415 lbs/ft. at 4000, to 465 at the same rpm. We asked one Chevrolet man why the increase in displacement and torque didn't have a parallel increase in horsepower. "This was done primarily to save weight," he said, with a twinkling eye; "you must remember that cast-iron is very heavy, and by removing thirty cubic inches of it, we have made a significant reduction in weight."

Last year's 396 "porcupine head" Corvette was cranking out quite a bit more than its advertised 425 bhp, and with 427 cu. in., the gap between advertised and actual becomes even broader. However, Chevrolet insists that there are only 425 horses in there, and we'll just have to take their word for it. Though we feel compelled to point out that these are 425 horses of a size and strength never before seen by man—horses as tall as houses, with hooves as big as bushel baskets. When you have *this* many of *those* horses exerting their full force against the small of your back, you are profoundly impressed, and you will most likely lose all interest in counting anyway.

Last month we said that the most dominant feature

of any Ferrari was its engine. Well the same thing is true of this big 427 Sting Ray, except that we'd go one step farther and say that it's the *power,* more than the engine, that overwhelms every other sensation. There's power literally everywhere, great gobs of steam-locomotive, earth-moving torque. And because the Corvette is a relatively small object, and considerably lighter than the other American production cars that are fitted with such engines, there's a direct, one-to-one relationship between the amount of throttle opening and the physical and emotional sensations the driver feels.

Other Corvettes, either the current models with 327 cu. in. engines, or the older ones with the original 283, had a more "European feel." They were very powerful, but they had a zippy, high-winding, no-flywheel feel to them that's missing from the big 427. According to Zora Arkus-Duntov, Chevrolet's Corvette specialist from the very beginning, this difference in sensation is directly related to the comparative torque curves of the two engines. The hottest version of the 327, the 375-bhp fuel-injection engine, produced 350 lbs/ft of torque at 4600 rpm. The new 427 produces 465 lbs/ft at 4000 rpm. As you can see, the 327's *rate* of acceleration was still increasing after the 427's torque curve had dropped off. Thus, even though the 427 is delivering more torque and more acceleration, its earlier torque peak results in less *sensation* of acceleration in the higher rpm ranges.

Except for compensatory suspension changes and a few other twists and improvements, the 427 Sting Ray is basically identical to any other Sting Ray. But the big jump in power that came with the jump from 327 to 427 makes those detail changes critical.

Higher-rate front and rear springs are fitted, along with a ⅞-inch stabilizer bar at the front and a ¾-inch bar at the rear. The half-shafts and U-joints are shot-peened, and are constructed of stronger stuff than on the 327, and there's been in increase in both coolant and lubricant capacity. The 427 can only be ordered with the "Muncie" four-speed transmission and Chevrolet's "Positraction" limited-slip differential. All prudent precautions, which underline the massive thrust available from this new powerplant.

In addition to our 425-bhp test car, Chevrolet also offers a 390-bhp version of the 427-cu. in. engine. This engine has different heads with smaller intake ports, and a smaller-section intake manifold which is cast-iron, as opposed to aluminum on the 425. It also has hydraulic lifters instead of the mechanical lifters in the hotter engine. Both have single four-barrel Holley carburetors, and there's no difference in exterior appearance, save for various emblems and decals and things. The hydraulic-lifter engine develops 390 bhp at 5400 rpm, and 465 lbs/ft of torque at 3600. It's not what you'd call puny, by any stretch of the imagination. In fact, without a watch or a measured quarter-mile, the average driver would have a hard time telling the difference—except for a smoother idle and less mechanical noise.

The interior appearance has no significant changes either. It is still roomy, comfortable, and very well sealed against wind and weather. The seats have a broad range of adjustment, and though they're not the super-buckets of a Ferrari GTB, they are pleasant to sit in for long drives. The back rests still are not adjustable, and this is too bad, especially now that virtually every medium-to-high-priced sedan in the GM line-up offers that feature.

The driving controls and small switches are just fine. The steering wheel telescopes in and out, and this, combined with the seat adjustment, makes it possible to have a pretty decent driving position. Our test car had power brakes and power steering, and we were grateful for both. GM's Saginaw Steering Gear Division has worked hard to provide a normal American-style power steering gear that has some feel and accuracy, and they've done a good job. The Mercedes-Benz system was their performance-target and though the Corvette doesn't measure up quite to that standard (you're still turning a valve, instead of the wheels), it *is* superior to any other American car. The power brakes are even better. They're smooth, free from any grabbiness or directional instability, and they do get it stopped! Even bringing it down from speeds in the 130-140 bracket without any sweat.

Driving the car on an unrestricted proving ground road is a memorable experience. It accelerates from zero to 100 in less than eleven seconds—faster than a lot of very acceptable cars can get to sixty—and is so smooth and controllable in the three-figure speed ranges that it all becomes sort of unreal. In fact, in those circumstances it's pretty hard to tell anything about the car at all, except that it goes like bloody hell and stops and steers without scaring you.

Everything comes into focus when you get it out on the public roads. Compared to anything you might come up against—unless you're unlucky enough to encounter a Cobra 427 (see page 37)—it's the wildest, hottest set-up going. With the normal 3.36 rear axle ratio it'll turn a quarter mile that'll give a GTO morning sickness, and *still* run a top speed of around 150 mph. The 327-engined version is still our favorite, but if you must go faster than anybody else, and you insist upon being comfortable, this is a pretty wild way to go.

The difference between this seven-liter street machine and all the big seven-liter super stocks is in size and proper suspension. All that murderous acceleration is balanced by excellent, almost-lightweight handling. It's stiff and stable and it gets the power on the road—when the wheels stop spinning. First gear is apt to be all wheelspin if you're not careful, and second is almost as bad. Even third can break the rear end loose, if you're down below the 4000-rpm torque-peak, and there's hardly a road in America where you won't be—since that's the equivalent of something in excess of 70 mph! Judicious applications of throttle will eliminate most of the spinning, but there just isn't any way to avoid it in first.

The extra weight of this big engine doesn't really seem to affect the car's handling at all. There's a general feeling of ponderousness that one associates with any of the bigger sports machines at low speeds, but when you're going fast it's quick and responsive. It *is* more difficult to accurately place it in a fast corner, but this is more due to its power steering than to its bulk.

There's no sense even trying to make MGB or Porsche comparisons with this Corvette, because . . . well, because it's so uniquely *American*. It weighs over 3300 pounds but it'll do 140 before most European sports machines are out of second. It does not mince tidily around corners, but it gets around corners faster than most of its peers. It's an *American* GT car, and it'll hold its own in any company at any price.

The most fascinating thing about the continuing success of the Corvette, and the powerful appeal of this new seven-liter contrivance, is its relationship to the widely-publicized GM ban on racing. Many people thought that the Corvette would wither and die when it was no longer the car to beat in the races. To the contrary, they're selling better than ever—and going better than ever as well. It's a shame really, if they keep building them this good and this fast, they may *never* have to go racing again.

c/D

CORVETTE STING RAY 427

Manufacturer: Chevrolet Motor Division
General Motors Corporation
Detroit 2, Michigan
Price as tested: ca. $6000

ACCELERATION

Zero to	Seconds
30 mph	3.2
40 mph	3.8
50 mph	4.6
60 mph	5.4
70 mph	6.9
80 mph	7.8
90 mph	9.0
100 mph	10.6
Standing ¼-mile	112 mph in 12.8

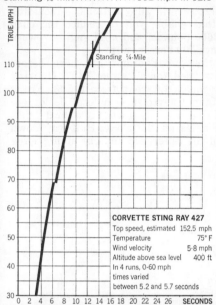

CORVETTE STING RAY 427
Top speed, estimated 152.5 mph
Temperature 75° F
Wind velocity 5-8 mph
Altitude above sea level 400 ft
In 4 runs, 0-60 mph times varied between 5.2 and 5.7 seconds

ENGINE

Water-cooled V-8, cast iron block, 5 main bearings
Bore x stroke..... 4.25 x 3.76 in, 106 x 94 mm
Displacement................ 427 cu in, 7000 cc
Compression ratio................ 11.0 to one
Carburetion................ Single 4-bbl Holley
Valve gear........ Pushrod-operated overhead valves, mechanical lifters
Power (SAE)........ 425 bhp @ 5600 rpm
Torque........ 460 lbs/ft @ 4000 rpm
Specific power output..... 0.995 bhp per cu in, 60.7 bhp per liter
Usable range of engine speeds. 500-6500 rpm
Electrical system... 12-volt, 61 amp-hr battery, 37A alternator
Fuel recommended.................... Premium
Mileage..................... 10-15 mpg
Range on 20-gallon tank........ 200-300 miles

DRIVE TRAIN

Clutch................ 10.5-inch single dry plate
Transmission................ 4-speed, all synchro

Gear	Ratio	Overall	mph/1000 rpm	Max mph
Rev	2.26	7.59	−10.3	−66
1st	2.20	7.39	10.7	69
2nd	1.64	5.51	14.3	95
3rd	1.27	4.27	18.5	120
4th	1.00	3.36	23.5	152.5

Final drive ratio.................... 3.36 to one

CHASSIS

Wheelbase................................ 98 in
Track................ F 56.8, R 57.6 in
Length................................ 175 in
Width................................. 69.6 in
Height................................ 49.8 in
Ground Clearance..................... 7.0 in
Dry weight........................... 3005 lbs
Curb weight.......................... 3160 lbs
Test weight.......................... 3380 lbs
Weight distribution front/rear........ 47/53%
Pounds per bhp (test weight)............ 7.9
Suspension F: Ind., unequal length wishbones, coil springs, anti-sway bar.
R: Ind., lower transverse link, half-shafts acting as upper locating members, trailing arms, transverse leaf spring.
Brakes.. 11.75-in discs, front and rear, 461.2 sq in swept area
Steering.................... Recirculating ball
Turns, lock to lock...................... 3
Turning circle........................ 36 ft
Tires............................ 7.75 x 15
Revs per mile........................... 760

CHECK LIST

ENGINE
Starting.................... Good
Response................... Good
Noise...................... Fair
Vibration.................. Fair

DRIVE TRAIN
Clutch action................. Excellent
Transmission linkage........... Excellent
Synchromesh action............. Excellent
Power-to-ground transmission....... Fair

BRAKES
Response..................... Excellent
Pedal pressure............... Excellent
Fade resistance.............. Excellent
Smoothness................... Excellent
Directional stability........ Excellent

STEERING
Response..................... Good
Accuracy..................... Fair
Feedback..................... Fair
Road feel.................... Good

SUSPENSION
Harshness control............ Good
Roll stiffness............... Good
Tracking..................... Good
Pitch control................ Good
Shock damping................ Good

CONTROLS
Location..................... Good
Relationship................. Good
Small controls............... Good

INTERIOR
Visibility................... Good
Instrumentation.............. Excellent
Lighting..................... Good
Entry/exit................... Fair
Front seating comfort........ Good
Front seating room........... Good
Rear seating comfort......... —
Rear seating room............ —
Storage space................ Poor
Wind noise................... Fair
Road noise................... Fair

WEATHER PROTECTION
Heater....................... Excellent
Defroster.................... Excellent
Ventilation.................. Good
Weather sealing.............. Good
Windshield wiper action...... Good

QUALITY CONTROL
Materials, exterior.......... Good
Materials, interior.......... Good
Exterior finish.............. Good
Interior finish.............. Good
Hardware and trim............ Good

GENERAL
Service accessibility........ Good
Luggage space................ Poor
Bumper protection............ Good
Exterior lighting............ Good
Resistance to crosswinds..... Good

ROAD TEST 18/65

Seven liters can make a ton-and-a-half defy the laws of inertia!

427 Sting Ray

MAYBE IT'S DIFFICULT TO FEEL SORRY FOR A CORPORATION THE SIZE OF GENERAL MOTORS, but they certainly deserve sympathy for the "Nervous Nellie" routine they've been forced into since withdrawing from racing. What was supposed to fill these pages was a track test of an optioned 427, a hundred of which are supposed to be built by January One and homologated with both FIA and SCCA. Since their retirement from active competition, Chevrolet has been beseiged by customers and dealers to at least make hardware available so they could go racing on their own. This program was strictly that; give the customer something to work with. The package is still in the mill at this writing and MAY see a production line. If rumors are correct, it's nothing too wild; several alloy options — including cylinder heads — that enable you to reduce the curb weight to around 2600 pounds, and mild engine rework that puts output above 500 horsepower.

Anyhow, the optioned car didn't get here before our deadline so we settled for a "street" 427, figuring it wouldn't be very much different from a 396. Frank Milne, of Harry Mann Chevrolet — a local dealer enthusiastic about Corvettes — loaned us a brand-new unit with only 150 miles on the clock. Half-way down the block we noticed the difference and our staff photog

was already complaining of whiplash. It's no secret that the Sting Ray is a heavy car, heavier than almost any sports car on the market. Even the 396 didn't take away the heavy "feel" of driving a Sting Ray, but the 427 does. Oh, brother, it does. We didn't even bother taking it to Riverside or Willow Springs for acceleration runs... all you need is a two-block-long straight. From a 70-mph cruising speed you can accelerate to the redline in top gear (140 mph) in roughly a mile. Sixty to 100 mph in top gear takes a mere 7.2 seconds. Tell us you'd like a hotter performing road machine than this and we'll call you some kinda nut!

Before the test, Milne asked us if we'd like a special set of tires on the car. Knowing it takes quite a tire to absorb that kind of horsepower, we quickly agreed. What he put on were Firestone Sports Car "200's"; 6.70's on the front and 7.10's on the rear in combination with the six-inch $330 optional alloy wheels. They were great; as good as most in sandy-surface or wet-surface conditions, and sticking like mad in the dry. The only noticeable drawback was a very high-pitched whine at high cruising speed. Yet this was relatively quiet and more than offset by an almost total absence of cornering squeal. Undoubtedly they had a considerable amount to do with the excellent roadability of the car. There was little wheel-spin from a standing start, excellent stopping stability under extreme braking, and cornering power in the One G Category. The 200's are available from Firestone, and Harry Mann sells them for $205 extra on an exchange basis, installed.

Unexpectedly, the 427 is not a strong understeerer when there's any bite to be had. It is *slightly* understeering going into a corner, and this changes to slight oversteer as soon as power is applied. The limit is reached with the rear end breaking loose all of a sudden and, even though it takes quick reflexes, a slight twitch of the wheel and an instant of lift with the accelerator foot will bring it back under control. But you have to be going gawdawful fast to reach the limit—very near competitive racing speeds.

The porcupine engine... which first saw the light of day as a 427, *not* a 396... is a beaut. There's gobs of low end torque and a willingness to grab revs that belies its size. It'll turn seven grand, so the 6500 redline is conservative. Making our top speed runs, we found it wanted to pull well past the redline and, with a 4.11 final-drive, 4000 rpm equaled 80 mph, making the speed of our test vehicle a very real 140 plus. With the optional close-ratio gearbox, you can grab a fabulous zero-60-mph time; it's first cog all the way. The four-speed, full-synchro box is the Muncie unit. It shifts firmly and positively, but has a little tendency to grunch in First and Reverse when it's hot. The center-pivot floats of the Holley carburetor completely eliminate any dumping or flat spots in the turns, but there is a definite hot-start problem, due apparently, to percolation. You have to hold the throttle on the floor and crank for a while to get it to fire up under these conditions.

While we're on the subject of temperature, it's interesting to note that, while our test car was not equipped with any power options, it ran a steady water temperature of 212F in an ambient of 75°F. This is worrisome—until you get in traffic with it and find it doesn't climb any higher. The cooling system is pressurized to 15 pounds, has a copper radiator, and—obviously—should have glycol instead of water at all times. As far as effect of the engine is concerned, modern V-8's with thin-wall castings operate fine at these temperatures. Heat in the cockpit is noticeable. The smooth-flowing cast-iron headers and twin exhaust systems that run beneath the chassis make excellent radiators that even the thick fiberglass can't insulate in warm weather. It would seem almost a must to have the whole system removed, sand-

Like the 396, the 427 is easily distinguished by the large and pointed hood bubble. As shown here, the powerful car really takes the corners.

Viewed from the rear, there's little different about the '66 Sting Ray except the ventilators missing from the quarter posts. Up front, however, is a very tasteful egg-crate grille replacing the previous design.

Top left shows the neat and very efficient instrumentation along with the simulated-wood steering wheel. At right, the wide, flat Firestones — in conjunction with the optional cast wheels — are ideal for this application, with only slight drawbacks in noise level. Behind them are huge disc brakes that really stop the heavy car from the high speeds it can reach. Below, the 427-inch Chevy takes up a lot of room, though there were no power options to contend with. Lower right, the new Holley carb, with center-pivot floats is the only fuel system available with the "porcupine" powerplants. Copper-core radiator is used with these engines.

blasted, and liberally coated with Sperex VHT paint.

Body changes to the 427 include the hood bubble that appeared with the 396 engine option, a very tasteful egg-crate grille, and the removal of the vents in the rear quarter of the roof. Not having an earlier 'Ray to compare it to, we suspect the seats have been raised, a bit, as we no longer had that "buried-below-the-dash" feeling we remembered in previous Corvette tests. Forward visibility is excellent, but we immediately missed the adjustable steering column option, as the wheel is a bit too close for personal tastes. Genuine carp department: that simulated-wood steering wheel is attractive and just the right size, but let your hands get the slightest bit damp, or greasy, and it becomes the slipperiest thing you've ever grabbed. As steering effort is fairly high—with no power option — driving gloves are a must.

With the optional suspension, the ride is *very* firm, pleasant enough for smooth roads, but almost uncomfortable on bumpy, wavey surfaces. Yet we logged almost 350 miles in one day of covering all kinds of roads, and neither driver nor passenger felt any overall discomfort. As a matter of fact, the firmness represented security as the right foot kept getting heavier and heavier.

Gas mileage is not the forté of the 427 Sting Ray. We found it easy to empty the 20-gallon tank of premium fuel; something in the neighborhood of 14 mpg. Conservatively driven, this figure would probably improve to 16 or 17 mpg, but we weren't in the mood for feather-footing with this kind of power and handling on tap. Incidentally, either the timing was slightly off or the 11-to-1 compression ratio is critical to detonation. It is advisable to use top-octane fuel, in the neighborhood of 105 Research. Our test car was equipped with transistorized ignition and, of course, the usual radio shielding required in a fiberglass car. There were no other spark problems. The $199.00 AM/FM radio worked beautifully. Lord, it ought to for that money! Seriously, it pulled in strong signals in areas we never believed it possible, with mountain ranges being no obstacle.

Other than the above, the new Sting Ray is pretty much like last year's. The bodywork gets better every year, and it's really first-quality this go-around. The base price of the 427 Ray is $4290; a pretty amazing figure for a high-performance limited-production machine. The options, like CR gearbox, engine goodies (that actually raise the output to 450 hp at something like 6200, but is underestimated to keep a couple of ambitious Eastern Senators from getting excited), tinted windshield, power windows, radio, etc., raise the list of this particular unit to $5400 delivered in Los Angeles. Add another six hundred odd bucks for the special tires and wheels — they're well worth the investment — and the total price is still under anything on the market that will match any of its individual attributes, let alone the sum of them. It's a really fine Grand Touring machine and the first Corvette we could get excited about owning!

— *Jerry Titus*

CHEVROLET STING RAY 427 SPORT COUPE — ROAD TEST 18/65

Vehicle	Chevrolet Sting Ray
Model	427 Sport Coupe
Price (as tested)	$5396.85 FOB LA
Options	A whole bunch

ENGINE:
- Type .... 8-cylinder V, water-cooled, cast-iron block
- Head .... Removable, iron
- Valves .... OHV, pushrod/rocker actuated, solid lifters
- Max. bhp .... 425 @ 5600 rpm
- Max. Torque .... 460 lbs. ft. @ 4000 rpm
- Bore .... 4.25 in. 108 mm.
- Stroke .... 3.76 in. 95.5 mm.
- Displacement .... 427 cu. in. 6994 cc.
- Compression Ratio .... 11.0 to 1
- Induction System .... Single center-pivot Holley, special alloy manifold
- Exhaust System .... Cast headers into 2 pipes and mufflers
- Electrical System .... 12V distributor ignition (shielded) alternator

CLUTCH:
- Single disc, dry
- Diameter .... 10.5 in.
- Actuation .... mechanical

TRANSMISSION:
- 4-speed, full-synchro, close ratio
- Ratios: 1st .... 2.20 to 1
- 2nd .... 1.64 to 1
- 3rd .... 1.27 to 1
- 4th .... 1.00 to 1

DIFFERENTIAL:
- Spring, with posi-traction option
- Ratio: .... 4.11 to 1
- Drive Axles (type): ... open, 2-joint, no slip coupling

STEERING:
- Worm & sector
- Turns Lock to Lock: 3¾
- Turn Circle: .... 37 ft.

BRAKES:
- Caliper discs, no power assist
- Drum or Disc Diameter .... 11.75 in.
- Swept Area .... 461.2 sq. in.

CHASSIS:
- Frame .... Tubular steel
- Body .... Fiberglass
- Front Suspension: ... Unequal arm, coil springs, tubular shocks, anti-sway bar
- Rear Suspension: .... I.R.S. transverse leaf spring, tube shocks
- Tire Size & Type: ... Firestone Sports Car "200" 6.70 x 14 front, 7.60 x 15 rear

WEIGHTS AND MEASURES:
- Wheelbase .... 98.0 in.
- Front Track: .... 56.8 in.
- Rear Track: .... 57.6 in.
- Overall Height .... 49.8 in.
- Overall Width .... 69.6 in.
- Overall Length .... 175.1 in.
- Ground Clearance .... 7.5 in.
- Curb Weight .... 3095 lbs.
- Test Weight .... 3420 lbs.
- Crankcase .... 5 qts.
- Cooling System .... 22 qts.
- Gas Tank .... 20 gals.

PERFORMANCE:
0-30	2.6 sec.	0-70	6.4 sec.
0-40	3.5 sec.	0-80	7.3 sec.
0-50	4.2 sec.	0-90	9.7 sec.
0-60	4.8 sec.	0-100	11.2 sec.

60-100 4th gear .... 7.2 sec.
Standing ¼ mile .... N.A. sec.
Top Speed (av. two-way run) .... 140 mph

Speed Error 30 40 50 60 70 80 90
Actual 30 40 50 60 71 81 91

Fuel Consumption
- Test: .... 14 mpg
- Average: .... 16 mpg

Recommended Shift Points
- Max. 1st .... 60 mph
- Max. 2nd .... 80 mph
- Max. 3rd .... 103 mph
- Max. 4th .... 140 mph

RPM Red-line .... 6500 rpm

Speed Ranges in gears:
- 1st .... 0 to 60 mph
- 2nd .... 5 to 80 mph
- 3rd .... 18 to 103 mph
- 4th .... 25 to top mph

Brake Test: .74 Average % G, over 10 stops. No fade encountered.

REFERENCE FACTORS:
- Bhp. per Cubic Inch 1.0
- Lbs. per bhp. 7.3
- Piston Speed @ Peak rpm .... 3509 ft./min.
- Sq. In. Swept Brake area per Lb. 0.149

IT'S GOT THE GO-NOW...
YOU NEED THE KNOW-HOW!

CORVETTE 427 ROAD TEST

by Bob McVay, *Associate Editor*

Wide mouth of gas-hungry Holley carburetor is kept clean with paper-element filter.

WE ONCE TRIED a young neighbor's motorized skateboard and found that it had two qualities in common with the 427 Corvette: They both go very fast, and both require a considerable measure of driving expertise.

Our test Corvette was equipped with the most brutish version of the biggest engine offered by a company that disclaims any interest in racing. It coyly rates it at 425 hp, but we think an extra two teams of Borax mules lie hidden behind the barn door. Engineers call it "porcupine," but that refers to the valve layout and not its agility.

The 427 has the kind of torque that made World War II fighter planes try to wrap themselves around their propeller on take-off. In the relatively light, front-end-heavy Corvette this verve tends to pave the highway with your rear-tire treads. With the standard 4-speed manual transmission and a little clutch diddling, the 427 is almost as adept as the old KB Lincoln in starting from stop in high gear. We don't recommend this, of course, and neither do we recommend starting off at more than quarter-throttle in low gear.

For drivers who have the guts and skill to master it—and the maturity to recognize it for what it is and handle it accordingly—the 427 Turbo-Jet Corvette is a road king. The other 99% of the population is likely to be much happier and safer in either of the two 327 models.

With a 3.08 rear end and at 6700 rpm, the 427 will theoretically be traveling at 170 mph. The 3.08 is a factory option and 6700 is only 200 rpm into the red-shaded warning area on the tach, so we will not argue with the chart that Chevrolet includes with its literature. All we know is that at 6500, we were doing about 135 on the back straight at Riverside and Turn 9 was coming up awfully fast.

On wet pavement, the power goes to the standard whitewalls but not to the road. Even though the steering will lose its ease, we don't think we'd own this car without changing over to Cinturato or Michelin-X (or their equivalent) tires, along with the optional 6-inch-rim aluminum wheels — anything to further a closer association between the 'Vette and the road, wet or dry.

In other respects, our convertible was at home in the rain. Unlike earlier models, it was completely weatherproof with a top that may be easily flicked up or down with one hand. Four-wheel disc brakes responded in their usual fashion, and although most descriptions fail to mention it, the Corvette people install tandem master cylinders. It is a braking system that could well be the model for the rest of the industry.

The instruments are honest, round,

(Top) Good-looking body is highlighted in daytime by absence of headlights, which recess behind lids at touch of a switch.

(Above) Only a brief glance is needed to read large, clearly marked gauges. Trip odometer and clock add to the dash array.

(Top) Folding soft top requires little effort to stow behind seat area. Added luggage space is afforded when the top is up.

(Above) Optional 17.6-to-1 ratio fast steering keeps driver in close control of front wheels through tight or long curves.

easy-to-read gauges. Cockpit ventilation problems have been eradicated, although we felt a little foolish using one hand to maintain our delicate contact with the road and the other to flick the electric windows up and down.

For drivers who don't want a car that demands complete concentration at all times, Corvette offers 300- and 350-hp versions of the well proven 327-inch V-8 with the 3-speed, all-synchromesh gearbox as standard and the 4-speed as an option. Powerglide (2-speed) is available only on the 300-hp model. The 427s come only with 4-speed transmissions, but owners can choose between a 2.20 or a 2.52-to-1 low-gear ratio, except on the 425-hp engine, which comes with the 2.20 ratio only. Both 427s have Positraction rear axles as standard equipment, with a choice of ratios from 3.08 through 3.36, 3.55, 3.70 and 4.11, to a low, low 4.56-to-1 rear end.

There are certain differences between the 427 engines. Both use extra-wide main-bearing caps, but the 425-hp job has 4 bolts holding each one on, while the 390 uses 2. With its mechanical lifters, the 425 has a slightly higher lift (thanks to a slightly wilder camshaft) than its much quieter 390-hp sister. Both sport huge hood bulges and special medallions, and both have stiffer front anti-roll bars and a similar device at the rear to help hold things in line.

As might be expected, gas mileage was not the 427's long suit — 8 mpg was our best in traffic, while about 12 was all we could squeeze out at low cruising speeds. Heavy-duty brakes, H-D suspension and H-D transmission are special options for the 427s, and we'd recomment them to every owner.

For those rare individuals who want, and can handle, its potential, the 427 Turbo-Jet is a red-hot machine, but if it gets away from you, don't say we didn't warn you. /MT

CORVETTE
2-door, 2-passenger convertible

ACCESSORY PRICE LIST
Engine options: 350 hp	$105.35
to *425 hp	181.20
Automatic transmission	194.85
*4-speed transmission	184.35
Limited-slip differential	42.15
*Heavy-duty suspension	36.90
*Whitewall tires	46.55
*Disc brakes	std.
Power brakes	42.15
Power steering	94.80
*Power windows	57.95
Radio AM	70.60
*Radio AM/FM	199.10
Air conditioning	412.90
Tinted glass	15.80
*Bucket seats	std.
*Adustable steering wheel	42.15
*Clock	std.
*Tachometer	std.
Head rests	42.15
*On test car	

MANUFACTURER'S SUGGESTED LIST PRICE: $4225.75 (incl. taxes, safety equip't & PCV device)
PRICE OF CAR TESTED: $5259.20 (incl. excise tax, delivery & get-ready charges, but not local tax & license)
MANUFACTURER'S WARRANTY: 24,000 miles/24 months

SPECIFICATIONS FROM MANUFACTURER

ENGINE IN TEST CAR: Ohv V-8
Bore and stroke: 4.25 x 3.76 ins.
Displacement: 427 cu. ins.
Advertised horsepower: 425 @ 5600 rpm
Max. torque: 460 lbs.-ft. @ 4000 rpm
Compression ratio: 11.0:1
Carburetion: 1 4-bbl.

TRANSMISSION TYPE & FINAL DRIVE RATIO: 4-speed manual, fully synchromeshed in all gears; 4.11 rear-axle ratio.

SUSPENSION: Independent front and fully independent rear; coil springs in front, semi-elliptic transverse in rear, tubular shocks at each wheel.

STEERING: Semi-reversible, recirculating ball nut
Turning diameter: 41.6 ft., curb to curb
Turns lock to lock: 2.92

WHEELS: Short-spoke disc, steel; 15-in. dia.

TIRES: 7.75 x 15 highway tubeless; rayon; 4-ply rated

BRAKES: Caliper disc, 4-wheel hydraulic
Diameter of disc: front, 11.75 ins., rear, 11.75 ins.

SERVICE:
Type of fuel recommended: Premium
Fuel capacity: 20 gals.
Oil capacity: 5 qts.; with filter, 6 qts.
Shortest lubrication interval: 3000 mi.
Oil- and filter-change interval: 3000 mi.

BODY & FRAME: Laminated fiberglass body; welded full-length frame with box construction
Wheelbase: 98.0 ins.
Track: front, 56.8 ins.; rear, 57.6 ins.
Overall: length, 175.1 ins.; width, 69.6 ins.; height, 49.8 ins.
Min. ground clearance: NA
Usable trunk capacity: 8.1 cu. ft.
Curb weight: 3360 lbs.

PERFORMANCE

ACCELERATION (2 aboard)
0-30 mph	2.5 secs.
0-50 mph	4.2 secs.
0-60 mph	5.6 secs.
0-75 mph	7.3 secs.

TIME & DISTANCE TO ATTAIN PASSING SPEEDS
40-60 mph	2.0 secs., 146 ft.
50-70 mph	2.0 secs., 172 ft.

STANDING-START QUARTER-MILE: 13.4 secs. and 105 mph

BEST SPEEDS IN GEARS @ SHIFT POINTS
1st	60 mph @ 6500 rpm
2nd	80 mph @ 6500 rpm
3rd	103 mph @ 6500 rpm
4th	135 mph @ 6500 rpm

MPH PER 1000 RPM: 20 mph
SPEEDOMETER ERROR AT 60 MPH: 10% fast
STOPPING DISTANCES: from 30 mph, 30 ft, from 60 mph, 120 ft.

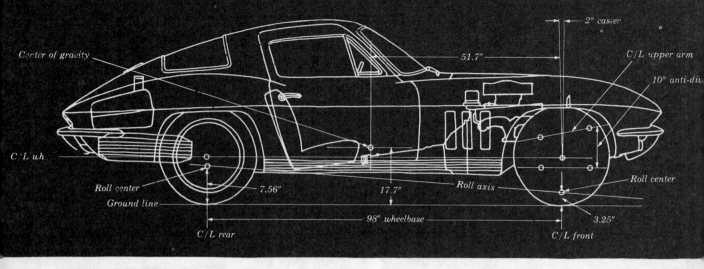

This kind of design gives Corvette surefooted handling characteristics.

I.R.S., unsprung weight and you

Some cars handle hard-traveling like your grandmother dispatching a garden snake with a rake. Lots of flailing and not much efficiency. That's because their suspensions were made for ordinary driving. But Corvette is no ordinary car. You expect it to have super handling, which is the reason for Corvette's surefooted independent rear suspension (I.R.S.)

There's nothing mystical about it. It all depends on those laws about mass and motion and such-like. For instance, most cars have a conventional rear axle. It's sturdy, simple and serves admirably for regular passenger cars. But when there's a bump, both wheels and the differential react together because they are a single mass. The high inertia involved makes it unsatisfactory for a maximum performance machine.

Corvette's differential is bolted to the frame independent of the wheels so it doesn't have to go up and down when the wheels do. This lowers the inertia and lets the wheels react quickly, which keeps the tires planted more solidly. That's called lowering the unsprung weight ratio. The Corvette suspension also allows each wheel to react to the road without affecting the other and with minimum camber change of its own so each tire maintains a firm grip.

Mount all this on variable-rate springs front and rear, add ten degrees anti-dive geometry and a very low roll center at the front, give it Corvette's carefully calculated rearward weight bias and you've got a machine that's really tidy in a hard corner. Which is one reason Corvette is still America's only true production sports car.

Handling—The Chevrolet Way

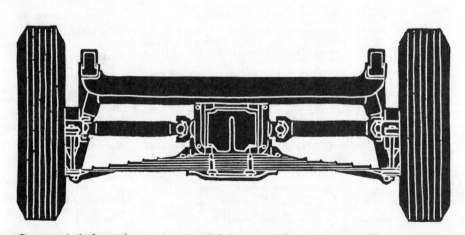

Corvette's independent rear suspension, revealed.

CAR LIFE ROAD TEST

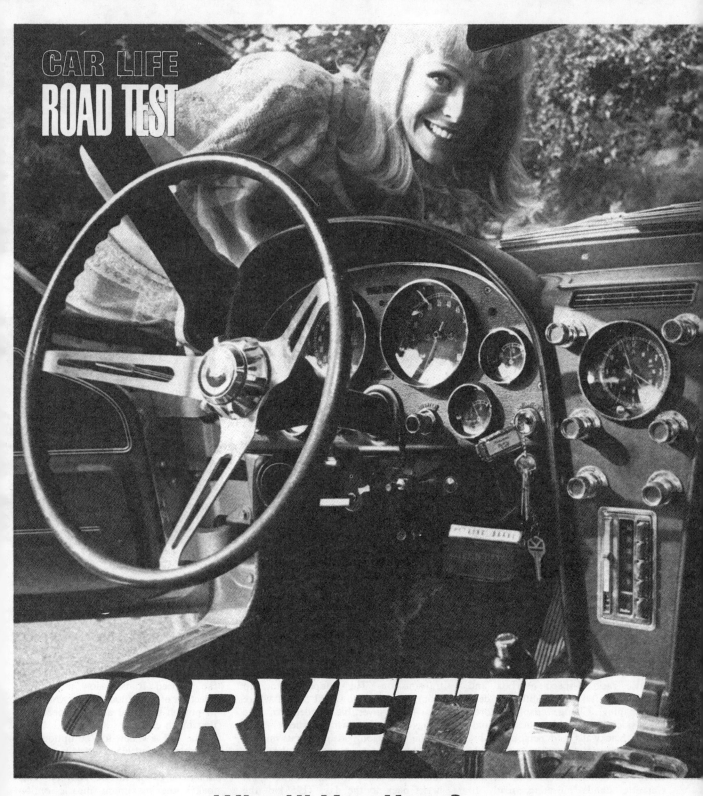

CORVETTES

What'll You Have?
427 cu. in. and 4-Speed, or 327 cu. in. and Automatic

THE ONE road test *CAR LIFE* staffmen most look forward to each year is that of the Chevrolet Corvette —this amiable, responsive vehicle never fails to delight its drivers and tickle its testers. It serves to renew their spirits and stir their blood just as a truly sporting car should do. And, of the dozen or so Corvettes *CL* testers have driven in the past half-decade, not a one has been disappointing.

For a car, particularly a car that changes as little as the Corvette, to retain this sort of mystique is enigmatic. Once tried, should not all Corvettes induce similar responses, similar reactions? As they spring from a common mold, should they not all be the same? Is a Corvette, a Corvette, a Corvette?

In truth, no two Corvettes can be

STING RAY fastback coupe has been a leader in a styling revolution.

HUSTLECAR 427/425 Corvette eats up dragstrips in 14-sec. gulps.

AUTOMATIC'S shift lever is in console along with air conditioner, radio controls.

CORVETTES

alike for only mere statistics and apparent lines resemble each other; each car tends to develop its own personality. Substitute an optional engine for a standard one and the car's character begins metamorphosis. Replace an automatic transmission with a close ratio 4-speed gearbox and the character takes on subtle but definite shading. Change crisply tailored white-sidewall tires for husky gold-striped blackwalls and personality begins to emerge; wrap the cockpit in a collapsible fabric top instead of a sleek, firm fastback cloak and the true Corvette stands pulsating, ready to obey its master's bidding.

In essence, those are the distinctions between *CAR LIFE*'s two test Corvettes. One is the sleek, torpedo-tailored towncar, the other a muscular, no-nonsense, do-it-right-now hustlecar. The difference in characters is accomplished by a wave of the option list and a flip of the computer's punchcard. The Corvette can be just what the doctor ordered, or it can be a psychotraumatic experience.

Naturally, the drive-line combination most affects the vehicle's personality. Where *CL*'s fastback Sting Ray Sports Coupe had the standard 327-cu. in./300-bhp and an automatic transmission, its Sting Ray Convertible had the top-option 427/425 with a 4-speed, close-ratio manual transmission. The coupe could slip along in silent, powerful surges where the 427 convertible tended to take things in great growling rushes.

The 427, along with 100 cu. in. and 125 bhp more, represents the very latest generation of engines for Chevrolet and the Corvette. The 327i harks back to the days (1955) of the 265-cu. in. Chevrolet V-8 although little but the basic proportions have remained the same. Light weight, great durability and relatively high output make the 327/300 a bargain in any Chevrolet, but particularly so in the Corvette. On test the 327 performed most adequately, making up in enthusiasm for what the 2-gear torque converter Powerglide lacked in versatility. With any transmission it would be a good performer.

However, a drive in the 427 can convince anyone with a drop of sporting blood in his veins that an overabundance of power can be controllable and greatly invigorating. There is nothing like turning on the valve of 425 bhp when the emphasis is on rapid departure from a fixed position. And, the 4-speed transmission makes the selection of how much power you want to put on the road at what speed just that much easier.

The 427 is a new-for-'66 option. An outgrowth of 1965's 396/425, it has its power peak at a lower rpm and a wider range of torque delivery. This makes the maximum muscle option more compatible to everyday driving yet doesn't seem to hamper its ultimate activity. Upper rev limits, however, are the same and this engine can safely turn 6500 rpm for brief spurts. Mechanical lifters facilitate this capacity and a big Holley carburetor, strong cam timing, special exhaust headers and a transistorized magnetic pulse ignition make it possible.

There are actually three transmission options for the Corvette. Standard is an all-synchromesh 3-speed (2.54

INSTALLATION of standard 327-cu. in. engine leaves enough extra space for air conditioner, emission-control pump and power accessories.

FILLING UP the engine space is no problem with 427—it uses up most of it in cast iron. Dual master cylinder and power booster are at left.

first, 1.50 second); the two 4-speeds also have synchromesh on all forward gears but vary in ratios—2.52, 1.88 and 1.46 for the wide ratio set, and 2.20, 1.64 and 1.27 for the closer set. The automatic has a single 1.76:1 intermediate gear plus a torque converter stall ratio of 2.1:1. The two 427 engine options include a 4-speed transmission (wide ratio with the 390-bhp version, close ratio with the 425) but Powerglide can only be ordered with the 327/350. A modified Powerglide was to have been included for the 427/390 at mid-year.

The reasoning behind the dropping of the old fuel injection 327 in favor of the 390/425 and subsequent 427/425 was obvious concern with price. Where the fuel injection engine cost the Corvette buyer an additional $600, the "full-house" 427 adds only $313 to his contract. Comparing performance results with earlier *CL* tests shows that more inches and reduced complexity have been beneficial. Where a 4.11:1-geared, fuel-injected '64 Corvette thrashed through the standing quarter-mile in 14.6 sec. with a terminal velocity of 98 mph, the test of the 390/425 (3.70 gears) last year got 14.1 sec. and 104 mph. The '66 model, with its bigger but more docile engine and "economy" gearing of 3.36:1, ripped off 14 sec. flat and 102 mph. With the optional 4.11s and some drag-slick tires, it could cut that down to 115 and the mid-13s.

The Corvette's great accelerative performance, while the most readily demonstrated, is not necessarily its most outstanding attribute. Rather, its maneuverability, braking and superb handling must share equal positions of exaltation. It takes high-speed travel over a variety of roads and through a combination of curve radii to appropriately reveal the car's inner beauty. In this context, either the 327/automatic or 427/4-speed can give its owner that sense of pride which comes from any superior creation.

The superior handling of the all-independently sprung Corvette chassis has to be experienced to be completely understood. And, once sampled, it makes a driver into a believer in this way of doing that job. That it does a superior job no longer can be doubted; that the independent system's advantages are not in wider use can only be decried. The same attitudes must apply to the 4-wheel disc brakes. Again, this has proved, both in development (see Page 50) and in actual usage, a superior way of doing the job. *CL*'s usual deceleration tests for brakes from 80 mph gave consistent, superior rates of stopping; only when we made a series of consecutive stops from 120 mph did demon fade rear his smelly head. Two of those were enough to produce a markedly spongy pedal on the 427. Consider, however, that these were the standard pads and discs, not the heavy-duty competition components which don't fade even under that sort of abuse.

As sophisticated as is the Corvette chassis, the two test cars presented something of a paradox. The lower-powered coupe was quiet, smooth and pleasant riding where the big-power convertible jerked and jounced its passengers and subjected them to unnecessary wind noise. The wind-catching differences between the smoothly fitted fiberglass coupe top and the overlapping-edged convertible top no doubt accounted for the greater noise level. The riding differences must be traced to the chassis itself.

Very few changes in basic chassis components are made between the two versions, the most significant one being a rear axle anti-roll bar with the 427-cu. in. engines. This link-type stabilizer is 0.562 in. in diameter and helps offset the understeering effect of their greater concentration of weight on the front wheels. The 427-equipped Corvettes also have a larger diameter front anti-roll bar, 0.875 vs. 0.75 in., so perhaps the cumulative effect of the two is what makes the 427s feel harsher. Ride rates at the wheels, usually good indicators of the relative firmness of spring action, are listed as the same figures—80 lb./in. front, 123 lb./in. rear.

This brings out another enigma of the Corvette chassis, its inability to transmit a feeling of structural integrity to its driver. Workings of the marvelous suspension systems, unequivocally the best produced in the U.S. today, are felt as separate bumps and thumps. Power applications and withdrawals are sensed as additional pulses. The driver senses every movement, every sound. In fact, the driver gets the impression he is sitting on the car, rather than being in, and an integral part of it. Perhaps this stems from the Corvette's separate, massive, ladder-type frame, a virtual bridge-truss which constitutes the strength of the car. The fiberglass body panels carry no structural loading, so the frame must do all the work. Whenever a

AIR-CONDITIONED and power-assisted, this Corvette is the torpedo-tailored towncar of docile alacrity.

vibration or bump strikes this big anvil, the driver, sitting on top of it, vibrates with the shock wave. It definitely lacks the all-of-a-piece feeling common to unitized body-chassis constructions.

Externally and internally, Corvette styling has remained little changed from 1963. The shark-like curvatures of the body panels are unchanged, though some of the gill openings have been modified. The vents just behind the coupe's doors were eliminated for '66. The 427-equipped models have a distinctive badge in their bulging hoods—the blister is necessary to provide adequate clearance for the carburetor air cleaner. Inside, control and instrument arrangements are identical to previous models. Both test cars were fitted with the optional telescopic steering column, a handy device where drivers of differing stature use the same car.

Finish of the exterior was characteristic of large molded fiberglass paneling; it was minutely rippled and inaccurately fitted. Apparently, this rippling is something Corvette owners always will have to accept. It has marked the Corvette from the very first and has shown little improvement in the intervening 12 years. The fit and finish of the interior, however, was as good as or better than previous models. Though the instrument panel layout still leaves something to be desired (particularly in the placement of the auxiliary, engine condition

1966 CHEVROLET
CORVETTE SPORT COUPE

DIMENSIONS
Wheelbase, in. 98.0
Track, f/r, in. 56.8/57.6
Overall length, in. 175.1
  width 69.2
  height 49.6
Front seat hip room, in. ... 2 x 21
  shoulder room 48.4
  head room 37.0
  pedal-seatback, max. 45.0
Rear seat hip room, in. ....
  shoulder room ....
  leg room ....
  head room ....
Door opening width, in. ... 31.0
Floor to ground height, in. . 8.3
Ground clearance, in. 5.0

PRICES
List, fob factory $4395
Equipped as tested 5573
Options included: Air cond., Positraction axle, emission control, tinted glass, am/fm radio, telescopic steering shaft, wsw tires, Powerglide, power windows, brakes & steering.

CAPACITIES
No. of passengers 2
Luggage space, cu. ft. 10.6
Fuel tank, gal. 20.0
Crankcase, qt. 5.0
Transmission/diff., pt. ... 3.0/3.7
Radiator coolant, qt. 19.0

CHASSIS/SUSPENSION
Frame type: Ladder, 5 crossmembers.
Front suspension type: Independent by s.l.a., coil springs, telescopic shock absorbers, link-type stabilizer.
  ride rate at wheel, lb./in. 80
  anti-roll bar dia., in. 0.75
Rear suspension type: Independent with lateral struts, U-jointed axle shafts, transverse leaf spring, telescopic shock absorbers.
  ride rate at wheel, lb/in. 123
Steering system: Linkage-assisted power, recirculating ball-nut, parallelogram linkage; spherical joint knuckles. Telescopically adjustable column.
  gear ratio 16.0
  overall ratio 17.6
  turns, lock to lock 2.9
  turning circle, ft. curb-curb .. 39.6
Curb weight, lb. 3210
Test weight 3620
Weight distribution, % f/r ... 52/48

BRAKES
Type: Dual-line hydraulic; caliper discs on radially vented rotors.
Front rotor, dia., in. 11.75
Rear rotor, dia. 11.75
total swept area, sq. in. 461.2
Power assist .... integral, vacuum
line psi @ 100 lb. pedal 500

WHEELS/TIRES
Wheel size .... 15 x 5.5K
  optional size available ..... 15 x 6L
  bolt no./circle dia., in. 5/4.75
Tires: B. F. Goodrich Silvertown 660
  size 7.75-15
  recommended inflation, psi 24
  capacity rating, total lb. ... 4400

ENGINE
Type, no. cyl. .... V-8, ohv
Bore x stroke, in. .... 4.00 x 3.25
Displacement, cu. in. 327
Compression ratio 10.25
Rated bhp @ rpm .... 300 @ 5000
  equivalent mph 115
Rated torque @ rpm .... 360 @ 3400
  equivalent mph 78
Carburetion .... Holley, 1x4
  barrel dia., pri./sec. 1.562/1.562
Valve operation: Hydraulic lifters, pushrods, overhead rockers.
  valve dia., int./exh. 1.94/1.50
  lift, int./exh. 0.399/0.399
  timing, deg. 32-87, 74-45
  duration, int./exh. 300/300
  opening overlap 78
Exhaust system: Dual, reverse flow mufflers.
  pipe dia., exh./tail 2.5/2.0
Lubrication pump type .... gear
  normal press. @ rpm ... 30 @ 1500
Electrical supply .... alternator
  ampere rating 37
Battery, plates/amp. rating ... 66/61

DRIVE-TRAIN
Transmission type: Torque converter with automatic planetary gearbox.
Gear ratio 4th ( ) overall .....
  3rd ( ) ....
  2nd (1.00) 3.36
  1st (1.76) 5.92
  1st x t.c. stall (2.10) .. 12.42
synchronous meshing? .... planetary
Shift lever location .... console
Differential type: Hypoid, overhung pinion, semi-floating axles.
  axle ratio 3.36

TAKE AWAY those whitewalls and add a folding top and the Sting Ray begins to look mean and purposeful.

instruments) it still presents the driver with complete information as to what is going on. Perhaps what was most appreciated was the big, round tachometer with green, yellow and red areas brightly marked off, stuck right up where the driver can see and utilize it.

The Corvette still does not have an outside trunk access, which may well be the only real design flaw in the whole concept. Entrance to the luggage area, which is ample on roadster and coupe bodies, is through the door and over the seat.

The biggest problem with the Corvette, however, has nothing to do with either its design or construction. Rather, it concerns the dealer-service situation after purchase of the car. Few Corvette buyers *CL* has encountered seem to have been satisfied by dealer treatment; most have some sort of complaint, whether or not justified. Generally, the complaint is simply that the dealer was far more interested in selling the car than in giving it proper and satisfactory service. "I felt as if I'd bought an imported car," was how one owner put it. Another had to take her '66 Corvette back to the shop a half-dozen times—each time to get things fixed from the mistreatment of the car during the last time the car was in for shopwork. Long delays for parts also have been reported to *CL*, along with dealer refusals to fix things because "they're all that way—there's nothing we can ▶

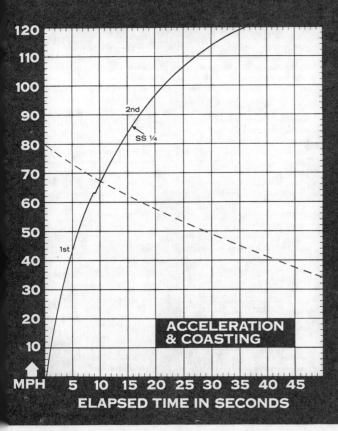

CAR LIFE ROAD TEST

CALCULATED DATA
Lb./bhp (test weight)	12.1
Cu. ft./ton mile	143
Mph/1000 rpm (high gear)	22.9
Engine revs/mile (60 mph)	2620
Piston travel, ft./mile	1420
Car Life wear index	37.2
Frontal area, sq. ft.	19.2
Box volume, cu. ft.	348

SPEEDOMETER ERROR
30 mph, actual	26.5
40 mph	37.5
50 mph	47.3
60 mph	57.9
70 mph	68.7
80 mph	78.8
90 mph	91.0

MAINTENANCE INTERVALS
Oil change, engine, miles	6000
trans./diff.	12,000/as req.
Oil filter change	6000
Air cleaner service, mo.	6
Chassis lubricaton	6000
Wheelbearing re-packing	as req.
Universal joint service	none
Coolant change, mo.	24

TUNE-UP DATA
Spark plugs	AC-44
gap, in.	0.033-0.038
Spark setting, deg./idle rpm.	6/500
cent. max. adv., deg./rpm.	30/5100
vac. max. adv., deg./in. Hg.	15/12
Breaker gap, in.	0.019
cam dwell angle	28-32
arm tension, oz.	19-23
Tappet clearance, int./exh.	0/0
Fuel pump pressure, psi.	5.25-6.50
Radiator cap relief press., psi.	15

PERFORMANCE
Top speed (5000), mph.	115
Shifts (rpm) @ mph	
3rd to 4th ( )	
2nd to 3rd ( )	
1st to 2nd (4850)	63

ACCELERATION
0-30 mph, sec.	3.4
0-40 mph	4.7
0-50 mph	6.4
0-60 mph	8.3
0-70 mph	10.8
0-80 mph	13.8
0-90 mph	17.3
0-100 mph	21.5
Standing ¼-mile, sec.	15.7
speed at end, mph.	86
Passing, 30-70 mph, sec.	7.4

BRAKING
(Maximum deceleration rate achieved from 80 mph)
1st stop, ft./sec./sec.	28
fade evident?	no
2nd stop, ft./sec./sec.	28
fade evident?	no

FUEL CONSUMPTION
Test conditions, mpg.	13.8
Normal cond., mpg.	14-16
Cruising range, miles.	280-320

GRADABILITY
4th, % grade @ mph.	
3rd.	
2nd.	18 @ 72
1st.	30 @ 51

DRAG FACTOR
Total drag @ 60 mph, lb.	150

do about it." Satisfied customer? No.

Doubtlessly there are two sides to the coin. However, the fact remains that these people are buying sophisticated, expensive cars and deserve reasonably satisfactory service for them. True, these people are usually people who like cars for their technical and/or romantic attributes and the purchase of a Corvette is more emotionally involved than the purchase of a more mundane vehicle for mere transportation. Thus they probably are more pernickety when it comes to their cars.

What's ahead for the Corvette in the face of the proliferation of the small, 4-seater sporting type of car? The forthcoming "Panther" will be in the same showrooms, and will compete in some ways, but will be lower priced, not as luxurious in concept or appointment, and not nearly as pleasing to the automotive enthusiast. *CL* thinks it will only serve to whet the appetite for the real thing. Corvette sales have climbed steadily even in the Mustang sales boom.

What will the Corvette of the future look like? Take a good look at the Mako Shark exhibition/experimental car being shown by General Motors. It could soon replace the Sting Ray. ■

Developing the Discs

IT WOULD BE difficult to pinpoint the exact moment when Chevrolet Division research engineers began working on the idea of disc brakes for Corvettes. The practical, production-line application of disc brakes to sporting-type cars had been around longer than the Corvette itself. Doubtlessly, the successful use of discs on European sports/racing cars implemented their thoughts, particularly those of Zora Arkus-Duntov whose specific responsibility has been development of the Corvette chassis. Himself a sports/racing driver of note on those same European race courses, Arkus-Duntov directed several special Corvette competition projects before the 1963 Sting Ray Grand Sport effort was undertaken.

The Grand Sport, it will be remembered, was to be a limited series of special, lightweight, purely competition-oriented Corvettes capable of achieving worldwide winners' circles. Development, testing and construction was well under way when General Motors' famous no-racing edict stopped the project early in 1964. A good portion of that early development and testing had been devoted to braking systems, in particular the disc brake.

Corvettes, at that time, already boasted one of the finest competition brake systems in existence—a $600 option which put heavy-duty, sintered-iron linings inside special finned drums with forced draft ventilation. These duo-servo shoes required a fairly high operating temperature for stabilized braking and the resultingly high hydraulic pressure requirement sometimes caused cracked drums. However, in terms of heat capacity and lining area, these were the best available for a car of the Corvette's weight and speed potential. Thus, the Sebring 12-Hour Race type of usage (maximum decelerations from 150 to 25 mph) was established as the standard.

According to Arkus-Duntov, one of the major reasons for disc braking is for its better modulation; i.e., the ratio of stopping power vs. pedal pressure required. "Duo-servo brakes do not lend themselves to good modulation," Duntov wrote in an SAE paper

1966 CHEVROLET
CORVETTE CONVERTIBLE

DIMENSIONS
Wheelbase, in.	98.0
Track, f/r, in.	56.8/57.6
Overall length, in.	175.1
width	69.6
height	49.8
Front seat hip room, in.	2 x 21
shoulder room	48.4
head room	38.5
pedal-seatback, max.	45.0
Rear seat hip room, in.	
shoulder room	
leg room	
head room	
Door opening width, in.	31.0
Floor to ground height, in.	8.3
Ground clearance, in.	5.0

PRICES
List, fob factory	$4295
Equipped as tested	5401
Options included: 427/425 V-8, Positraction, tinted windshield, trans. ignition, am/fm radio, telescopic steering shaft, gold stripe nylon tires, 4-speed close ratio, power brakes, steering and windows.	

CAPACITIES
No. of passengers	2
Luggage space, cu. ft.	8.1
Fuel tank, gal.	20.0
Crankcase, qt.	5.0
Transmission/diff., pt.	2.5/3.7
Radiator coolant, qt.	22.0

CHASSIS/SUSPENSION
Frame type: Ladder, 5 crossmembers.
Front suspension type: Independent by s.l.a., coil springs, telescopic shock absorbers, link-type stabilizer.
ride rate at wheel, lb./in. 80
anti-roll bar dia., in. 0.875
Rear suspension type: Independent with lateral struts, U-jointed axle shafts, transverse leaf spring, telescopic shock absorbers, 0.562-in. stabilizer.
ride rate at wheel, lb./in. 123
Steering system: Linkage-assisted power, recirculating ball-nut, parallelogram linkage; spherical joint knuckles. Telescopic column.
gear ratio 16.0
overall ratio 17.6
turns, lock to lock 2.9
turning circle, ft. curb-curb ..39.6
Curb weight, lb. 3270
Test weight 3610
Weight distribution, % f/r ..52.4/47.6

BRAKES
Type: Dual-line hydraulic; caliper discs on radially vented rotors.
Front rotor, dia., in. 11.75
Rear rotor, dia. 11.75
total swept area, sq. in. 461.2
Power assist ......integral, vacuum
line psi @ 100 lb. pedal 500

WHEELS/TIRES
Wheel size ....15 x 5.5K
optional size available ....15 x 6L
bolt no./circle dia., in. 5/4.75
Tires: UniRoyal Laredo
size 7.75-15
recommended inflation, psi 24
capacity rating, total lb. 4400

ENGINE
Type, no. cyl. ....V-8, ohv
Bore x stroke, in. ....4.25 x 3.76
Displacement, cu. in. 427
Compression ratio 11.0
Rated bhp @ rpm ......425 @ 5600
equivalent mph 128
Rated torque @ rpm ...460 @ 4000
equivalent mph 92
Carburetion ....Holley, 1x4
barrel dia., pri./sec. ...1.686/1.686
Valve operation: Mechanical lifters, pushrods, overhead rockers.
valve dia., int./exh. 2.19/1.72
lift, int./exh. 0.5197/0.5197
timing, deg. 54-102, 102-54
duration, int./exh. 336/336
opening overlap 108
Exhaust system: Dual, reverse flow mufflers.
pipe dia., exh./tail 2.5/2.0
Lubrication pump type ....gear
normal press. @ rpm ..50 @ 2000
Electrical supply ....alternator
ampere rating 37
Battery, plates/amp. rating ..66/61

DRIVE-TRAIN
Clutch type: Single disc, dry, centrifugal.
dia., in. 10.5
Transmission type: Manual, 4-speed
Gear ratio 4th (1.00) overall ...3.36
3rd (1.27) 4.27
2nd (1.64) 5.51
1st (2.20) 7.38
synchronous meshing ......all four
Shift lever location ....console
Differential type: Hypoid, overhung pinion; semi-floating axles.
axle ratio 3.36

on Corvette disc brake development. Without the inherent self-energization of the duo-servo brake, the disc system achieves a more linear relationship of stopping power to pedal effort. The driver simply gets only as much braking as his foot demands.

Virtually all early production types of disc brakes utilized a spot pad caliper clamping on a solid rotor. GM tests of these systems revealed a need for far greater capacity and durability if they were to be successfully adapted to the larger, heavier domestic product. Delco Moraine Division, which tested its first disc brakes in 1937, produced a vented disc in 1954 and subsequently installed 4-wheel sets on Buick, Oldsmobile, Cadillac and Corvette test cars with varying degrees of success. The radially vented rotor was first seen publicly on the Firebird II experimental car in 1955.

Calculations for the design objective of 1 G (32 ft./sec./sec.) deceleration showed that the higher heat rejection abilities of the vented disc had to be utilized for the Corvette Grand Sport. They also showed that the Corvette brake had to have the largest lining area compatible with two pistons per pad. Other design parameters included: Satisfactory operation without power assist, instantaneous pedal response, freedom from pad "knockback" by heavy cornering loads, and low sensitivity to weather and icing conditions. Unit cost, as always, was a major consideration.

Balancing of the 4-wheel system was based on 1 G braking conditions where 66% of the weight, because of forward inertia, was concentrated onto the front wheels. To achieve a 65/35 balance, front caliper pistons were specified at 1.875 in. diameter, the rear pistons at 1.375 in. Total piston (there were four per caliper) area was 33.9 sq. in., which produced a 43.2:1 hydraulic ratio and 196:1 overall ratio. This gave 0.005 in. of brake pad movement for every inch of pedal movement. And, by keeping pads in slight but constant contact with the rotors, the maximum caliper piston travel necessary was only 0.01 in. Thus total pedal movement was less with the discs than had been necessary with drums.

Keeping the pad in light contact was one of the more radical and origi-

CORVETTE DISC brakes work to perfection in competition, helped George Wintersteen drive to GT class victory in '66 Daytona Continental.

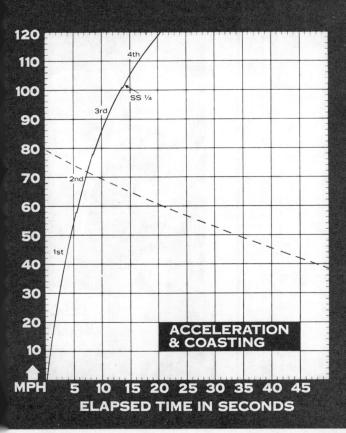

CAR LIFE ROAD TEST

ACCELERATION & COASTING

CALCULATED DATA
Lb./bhp (test weight)	8.5
Cu. ft./ton mile	180
Mph/1000 rpm (high gear)	22.9
Engine revs/mile (60 mph)	2620
Piston travel, ft./mile	1640
Car Life wear index	43.0
Frontal area, sq. ft.	19.3
Box volume, cu. ft.	351

SPEEDOMETER ERROR
30 mph, actual	30.1
40 mph	41.1
50 mph	51.7
60 mph	62.0
70 mph	72.2
80 mph	84.0
90 mph	94.8

MAINTENANCE INTERVALS
Oil change, engine, miles	6000
trans./diff.	as req.
Oil filter change	6000
Air cleaner service, mo.	6
Chassis lubrication	6000
Wheelbearing re-packing	as req.
Universal joint service	none
Coolant change, mo.	24

TUNE-UP DATA
Spark plugs	AC-43N
gap, in.	0.033-0.038
Spark setting, deg./idle rpm.	8/800
cent. max. adv., deg./rpm.	28/4600
vac. max. adv., deg./in. Hg.	15/12
Breaker gap, in.	(transistorized magnetic pulse)
cam dwell angle	
arm tension, oz.	
Tappet clearance, int./exh.	0.024/0.028
Fuel pump pressure, psi.	5.5-7.0
Radiator cap relief press., psi	15

PERFORMANCE
Top speed (5660), mph	130
Shifts (rpm) @ mph	
3rd to 4th (5600)	108
2nd to 3rd (5600)	78
1st to 2nd (6000)	58

ACCELERATION
0-30 mph, sec.	2.6
0-40 mph	3.5
0-50 mph	4.6
0-60 mph	5.7
0-70 mph	7.0
0-80 mph	8.7
0-90 mph	10.7
0-100 mph	13.4
Standing ¼-mile, sec.	14.0
speed at end, mph	102
Passing, 30-70 mph, sec.	3.4

BRAKING
(Maximum deceleration rate achieved from 80 mph)
1st stop, ft./sec./sec.	26
fade evident?	no
2nd stop, ft./sec./sec.	27
fade evident?	no

FUEL CONSUMPTION
Test conditions, mpg	11.6
Normal cond., mpg	11-13
Cruising range, miles	220-260

GRADABILITY
4th, % grade @ mph	20 @ 94
3rd	25 @ 81
2nd	33 @ 68
1st	off scale at 52

DRAG FACTOR
Total drag @ 60 mph, lb.	135

Developing the Discs

nal ideas in the development. It resulted in solutions to a good many of the operational problems previously thought insurmountable. The contact cost about 0.8 bhp at 100 mph—so little it could not be measured in normal fuel consumption tests—but in return it kept the rotor wiped clean of moisture and ice and pre-heated to an operating optimum (30-40° F above ambient air temperature). Pedal travel was reduced and the knock-back problem, where deflection of the disc by wheel distortion causes the piston to be pushed back into its cylinder, eliminated. Lining life, after 127 different compounds had been tried to find the best one, proved above expectations; tests showed 50,000 miles minimum.

Testing the Sting Ray's disc brake system was as unique as its development. Because of its capabilities, the system was immediately found to exceed all current braking standards, so, new standards had to be devised. The two major phases of testing were on-the-vehicle at the proving grounds, and laboratory dynamometer evaluation. The proving grounds checked the system as a whole, the laboratory exams were vital in selecting rotor configuration and pad material. The dynamometer schedule was developed to closely simulate both SAE and GM proving ground brake test programs; it accurately evaluated such things as fade, effectiveness, noise and wear; 173 programs were run on lining.

Proving ground tests always are the more dramatic. Yet the standard tests revealed no discernible fade in the fade-and-recovery section so a special abuse test was incorporated. This consisted of 20 stops from 100 mph at 1-mile intervals at a deceleration rate of 20 ft./sec./sec. (0.625 G). Fade and recovery characteristics proved far superior to the previously used drum brakes. Durability tests surpassed the Chevrolet standard of 36,000 miles. Dust tests, at GM's Mesa, Ariz., facility, consisted of 1700 miles at 50 mph with a stop every mile: Lining wear was slight and disc wear only 0.001 in. for the total distance. City traffic conditions were sampled in Phoenix and Los Angeles. The final acceptance came at Pikes Peak in Colorado. Runs were made from the summit (14,110 ft. elevation) to the gate house at the base of the peak with the transmission in neutral whenever braking was required. Full brake effectiveness was available at the end of each run, the SAE paper said.

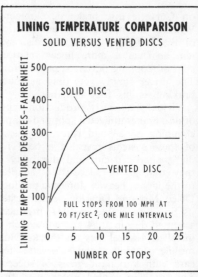

VENTED DISC gave significant reduction in lining temperatures during testing.

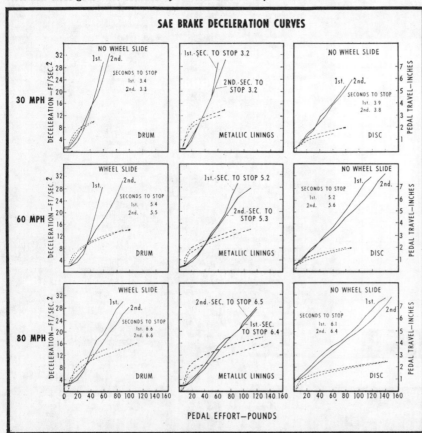

CURVES COMPARE results of tests with various Corvette brake configurations. Note how discs gave more consistency for least amount of pedal movement.

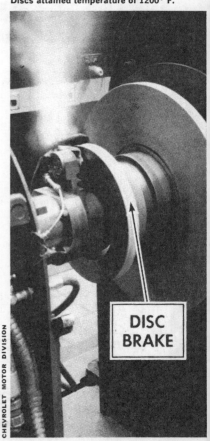

BRAKE DYNAMOMETER proved components. Discs attained temperature of 1200° F.

QUICKEST, FASTEST, STOPPINGEST 'VETTE YET!

BY MARTYN L. SCHORR

Unless you're wheelin' a Street Hemi or a 427 Cobra, steer clear of Chevy's hottest street stinger.

EVER SINCE it was endowed with a V-8 engine in 1955, the Corvette has enjoyed the honor of being one of *the* cars to beat on the street and strip. When the performance bug bit most of the manufacturers, starting in 1957, Corvette's position in the field was threatened. The Chevy boys hung in and increased the top power option from a 265 cube 195 hp eight to 283 cubes, 327 cubes, 396 cubes and finally in '66 a 427 cube affair. Even though a 427 engine is available in both solid and hydraulic lifter trim, the *old style* 327 wedge engine is still retained in standard and optional form. Corvette started out well ahead of the game (1955 Powerglide version would go to 60 mph in just 9.5 seconds) so they were never really involved in the cubic inch battle. They concentrated mainly on multi four barrel carburetion setups, solid lifter cams, big port heads, fuel injection and finally high riser single four barrel manifolds and "semi hemi" porcupine combustion chamber configurations in their quest for street-strip supremacy.

The latest offering from Corvette is a far cry from the original "plastic play toy" introduced in 1954. In top engine trim it is capable of accelerating to 60 mph in 5 seconds flat, red lining out at 140 mph and stopping in record time under the most adverse conditions. It can literally walk away and hide from any domestic production car, except for the Shelby 427 Cobra and MoPar street hemi.

Even though Chevrolet insists on being divorced from anything that even hints of racing, the 1966 Sting Ray can be ordered with a fantastic amount of genuine racing equipment. Both in the engine and handling departments. The standard engine this year is a rather mild mannered 300 hp 327 wedgie, which is not really bad for the country squire who wishes to use his Ray for rush hour commuting and around town transportation. Next up the line is a 350 hp 327 (RPO L-79) which sports a little more compression, a higher redline and a slightly wider torque range. That takes care of the standard 327 engine lineup which is basically a holdover from the days when the 327 was King of the Hill.

The really "hot setup" engine options are in the form of "semi-hemi" or porcupine chambered 427's which are direct descendants of the Daytona "Mystery" and Z-11 427 drag engines. Those engines were deep-sixed after the '63 Daytona fiasco. The "semi-hemi" was first made available

Semi-hemi 425 hp Ray can be factory fitted for road racing. Swingaway blinkers break up the flowing frontal lines.

Bubbled hood sets "semi-hemi" models off from conventional 327 wedgies.

Large matching speedo and tach are centered in the fully instrumented dash. Max acceleration figures were recorded getting off the line at 3400 rpm.

Above, rear deck flap hides the 20-gallon tank filler from view. Road racing 36-gallon fiberglass tank is available on special order. Right, high-riser Holley quad feeds the mix to the high lift cammed 425 hp screamer.

in the Sting Ray late last year in up to 425 hp 396 cube trim, after it had been announced as an option in full size Impalas and specially constructed Z-16 or Z-33 Chevelles.

In '66 the "semi-hemi" is available in two states of tune, with horsepower ratings of 390 and 425. The hydraulic-cammed 390 hp model (RPO L-36) sports a compression ratio of 10.25-to-1, cast aluminum alloy pistons, drop forged steel rods, a .461-inch lift cam, and a high riser Holley 3882835 four barrel with 1.562-inch primaries and secondaries. The 425 hp (L-72) model packs an 11.0-to-1 compression ratio, impact extruded pistons, high alloy steel rods, a .519-inch lift solid lifter cam, and a high riser Holley 3886101 four barrel with 1.686-inch primaries and secondaries. Besides the above mentioned goodies, the 425 hp version has larger intake valves, four-bolt main bearing caps instead of two-bolt models, and fittings for an externally-mounted oil cooler. If any type of competition is on the plans board, the 425 hp version is the only one to order in a new Sting Ray.

"You have just been stung by a Sting Ray"

Although not listed in the official Chevrolet AMA Sting Ray specifications book, there is another optional 427-425 hp engine which is even more suited for competition. All we know at this time is that this engine does exist and it features a gear-driven camshaft instead of the conventional chain driven setup. Before ordering a Sting Ray for all out competition it would be wise to check with a Chevy specialist such as Bill Thomas in Anaheim, California, or any dealer who engages in sports or drag competition on the availability of this "not listed" option.

Above & below, 11.75-inch power-assisted discs are used front and rear. Rear drum assembly incorporates conventional parking brake components.

With four wheel discs, HD suspension "Stinger" is right at home on any road course.

1966 CORVETTE STING RAY SPECIFICATIONS

ENGINE

Type	OHV V-8
Displacement	427 cubic inches
Compression Ratio	11.0-to-1
Carburetion	Single Holley Quad
Camshaft	Mechanical, .519-inch lift
Horsepower	425 @ 5600 rpm
Torque	460 foot/pounds @ 4000 rpm
Exhaust	Dual headers, dual pipes
Ignition	Dual points, transistorized

TRANSMISSION

Make	Muncie four-speed
Control	Floor shift

REAR END

Type	Semi-Floating Positraction
Ratio	3.70-to-1

BRAKES

Front	11.75-inch power-assisted discs
Rear	11.75-inch power-assisted discs

SUSPENSION

Front	Independent, HD coil springs, HD shocks, sway bar
Rear	Multi-leaf, torque control arms, independent
Steering	Manual
Overall Ratio	20-to-1

GENERAL

List Price	$4295
Price As Tested	$5400
Weight	3400 pounds
Wheelbase	98 inches
Overall Length	175 inches
Tire Size	7.75 x 15 Goodyear

PERFORMANCE

0 to 30 mph	2.8 seconds
0 to 60 mph	5.0 seconds
Standing ¼ mile	111 mph
Elapsed Time	12.80 seconds
Top Speed	140 mph (EST)
Fuel Consumption	10 mpg

Our test Sting Ray, a coupe painted electric blue, was factory equipped with an average amount of street accessories which included 3.70-to-1 positraction rear cogs. It was as loaded for bear as anyone could possibly want for the street. Factory rated with a curb weight of 3140 pounds, our test Ray tipped the Fairbanks at 3400 pounds. Starting with a base standard equipment list price of $4295, the actual list retail price jumped to just under $5400 when all the extras were added in.

On the street the 427 425 hp Sting Ray qualifies as just about the hottest thing on four wheels. The "semi-hemi" is chock full of so much torque that it requires little or no effort to plaster down two streaks of heavy black material in any gear at any time. During the timed acceleration runs we found that the engine craved for more than the 6500 revs that were redlined on the larged faced full sweep tach. All shifts were made over 7000 as there was no appreciable drop of torque when the tach needle soared past the red area on the tach. Before going on we should qualify the above statement. In stock out of the showroom form we were not able to coax much more than 6800 rpm before a power loss took place. Charlie Dodge and George Snizek at Pacers Auto spent many hours setting the valves, rejetting the center-pivot float Holley, building a curve into the transistorized distributor and trading the stock wires in for steel core wires (using the stock covering material). We made a survey and found that most of the 427's and late 396's on the road in stock dealer tune were not much quicker than 365 and 375 hp 327 '64 models.

After minor reworking the engine sounded and behaved like a custom built high performance machine. We ran some side by side acceleration tests and found the Sting Ray a fairly even match for a MoPar street hemi. In the handling and braking departments the Sting Ray out performed the full size Detroit monster by leaps and bounds.

With the 3.70 cogs and Muncie all synchro four speed (linkage adjusted by Pacers for a shorter, more positive throw) we were able to run to 60 mph in just 5 seconds flat, to 100 mph in 11.5 seconds and cover the quarter mile in less than 13 seconds. The four speed in our car was of the standard design and sported a 2.20-to-1 low gear as opposed to an optional close ratio unit (same optional cost as standard four speed) which features a 2.52-to-1 low.

The exhaust system can best be described as a throaty Hollywood setup and is one of two exhaust system options listed by Corvette. The stock reverse flow muffled system is quiet, respectable and easiest to live with. Our test car featured the Off the Road Conventional system ($37 extra) which means that crimped pipe replaces the standard mufflers. The other optional system is known as Off the Road, Side Mounted ($132 extra) and makes use of side-mounted crimped pipe with finned full length polished aluminum shields. This system places the noise right outside each door, tends to get to you after a while on long trips and is actually illegal in some states. Carroll Shelby found this out when he first marketed his GT-350 Mustangs with side-mounted exhaust outlets. You will note that his 1966 models have exhaust pipes that exit at the rear!

Matching the accelerating and top speed ability of the 425 hp powerhouse is a working suspension and chassis that makes the Sting Ray feel right at home on any handling course or skid pad. By mastering the oversteering and understeering characteristics of this high speed stormer, we were able to lay it into any corner and power it around. However, this cannot be done with a Sting Ray equipped with either the stock 7.75 x 15 inch whitewalls or high performance gold-stripes. Most of the tire manufacturers that engage in racing offer sticky shoes for this charger.

The stock fairly stiff front shocks, unequal arm coil spring front with an anti roll bar and independent rear with transverse leafs and stiff shocks are more than adequate for most driving conditions. However, for those buffs who require maximum stiffness and don't mind suffering when negotiating other than smooth paved roads, there is a special ultra-firm suspension available on 425 hp models.

Standard equipment on the Sting Ray, regardless of engine choice, is the finest braking system available on any domestic automobile. At the front and rear are 11.75-inch caliper disc brakes with cast iron vented rotors and on the firewall is a Delco-Moraine vacuum power assist unit ($42 extra). This flawless system affords the driver with 461 square inches of swept area and the peace of mind to cruise at the highest legal limits and to suffer through a lengthly series of panic stops. We tried more than ten attempts to fade the brakes (panic stopping from 75 mph) and were unsuccessful. A special built-in rear drum brake assembly permits the use of a conventional mechanical parking brake assembly.

The interior of our hardtop coupe was rather well laid out and is just the ticket for the single guy or girl who engages in limited touring. The stowage area behind the leather buckets ($79 extra-vinyl standard) topped off with headrests ($42 extra) is carpeted and can hold a reasonable amount of luggage. As can be seen in the interior photo, our fitted camera case, tripod and overstuffed gadget bag was rather lost in the deep area. Other extra cost options were gadget bag were rather lost in the deep area. Other extra cost options were the wood wheel ($47), AM-FM radio ($199), electric windows ($58). We found the seat belt buckle holders that are integral with the design of the console really boss and wished that the steering post had been fitted with the optional adjustable steering wheel. The stock position of the wheel placed it too close for maximum performance driving. We have nothing but praises for the well laid out, fully instrumented dash which we consider to be mandatory for any high performance car.

All in all we feel that the 425 hp Sting Ray is a fantastic performing machine that must be driven to be believed. It is not the car for every high performance buff, as handling it requires a certain degree of driving skill that only comes from driving a great variety of high speed machines. It handles differently from any other domestic machine and has its own little idiosyncrasies.

It's high cost is another factor which limits its acceptance among high performance buffs. Not only is the base list price of the coupe high, but the extra cost price of the accessories really places it in a class all by itself. The 425 hp engine option adds $313, the four speed adds $194 and if you want to go for knock-off aluminum finned wheels and a 36 gallon fiberglass fuel tank you can add an additional $516. As you can see the list can go way up to over $6000. But that's not really bad when you figure that you can't by a 427 Cobra, which doesn't even have roll up windows, for less than $7000.

1967 CORVETTE

Unique among American cars—and among sports cars

THE STING RAY is in its fifth and probably last year with that name and body style, and it finally looks the way we thought it should have in the first place. All the funny business—the fake vents, extraneous emblems and simulated-something-or-other wheel covers—is gone, and though some consider the basic shape overstyled, it looks more like a finished product now.

Actually, the car has undergone few changes in its 5-year run. The suspension and sound deadening were refined in the 2nd year, disc brakes introduced in the 3rd and the needlessly large "porcupine" engine added as an option in 1966. Minor refinements have been made all along; sales have risen gradually each year. And it remains unique among American cars —and among sports cars.

For this test we selected an almost basic version of the Corvette: a convertible hardtop with the standard 300-bhp 327 engine, 4-speed gearbox and power assists. With the exception of the power assists, we concluded that this is the Corvette for the thinking driver.

The test car was certainly impressive looking. Its paint was better than on any Corvette we've tested, a deep claret red. Its restrained trim and the new wide-rim (6 in.) wheels gave it a purposeful, almost elegant look.

Performance can be taken for granted with over 5 liters and 300 bhp in a 3160-lb car. It is a credit to the American way of doing things that the performance is achieved in a remarkably silent, smooth and economical manner. Our test car had the optional 4-speed gearbox, which is smooth, quiet and easy shifting (the linkage has been isolated from the vibration that was an annoyance in earlier models); but now that the standard 3-speed has been redesigned and includes a synchronized 1st, we can see no good reason for ordering the 4-speed for general use. Other than the ease of reselling, that is.

The only thing that marred the flexibility of performance in our test car was the unwillingness of its engine to return to idle speed. For one thing, its throttle linkage was a bit sticky; but more basic than that is the fact that the anti-backfire valve used with emission-controlling air injection dumps a blob of fresh air into the intake manifold for about 2 sec after the foot is lifted, causing the engine to hesitate momentarily at 1200 rpm on its way down to the normal 500-rpm idle speed. Minor, but bothersome, for California customers.

It's hard to find fault with the Corvette's handling; it's as near neutral as any car we know and of course there's always enough torque available to steer with the throttle. It is quite stable directionally in a cross wind and imparts an immense feeling of security out on the open road. Steering effort with the standard 20.2:1 overall ratio is moderate, and the customer may have the ratio lowered to 17.6:1 by a simple repositioning of the steering linkage and resetting of the toe-in, if he's willing to accept parking efforts in the Armstrong category.

Our test car, however, was equipped with the optional linkage power steering; and in our opinion this system is unacceptable. Like other GM power steering it is set to begin assisting the driver only after he has exerted a pre-set amount of force at the wheel rim. On the Corvette, this takeover-point apparently has been raised—most likely to give more "feel." But the result is almost the opposite. The wheel takes an almost constant amount of effort (about 7 lb) for any maneuver. Corvette is the last GM car to retain linkage-assist power steering and it's reasonable to predict that next year it will have the integral unit used on other GM models—and hopefully with more accuracy.

Over smooth roads the Corvette has a pleasing, almost big-car ride. Noise level is low from most sources, and thus this is a car for long-distance travel. Its effective heater, excellent AM/FM radio and good weather protection (regardless of top ordered) place it near the top of the list of sports/GT cars for pure comfort. On rough roads the ride deteriorates as the progressive-rate springs, limited wheel travel and willowy body structure combine to put up quite a fuss. Under these conditions it is no match for the best European examples, like the Mercedes 230 SL or BMW 2000 CS. But then this is More Car for the Money, and perhaps it is unrealistic to expect ultra-sophisticated chassis design. In all fairness, however, this is America's best chassis.

The Corvette's interior abounds with niceties. There are a handy new center-mounted handbrake, seat-belt hooks that solve the problem of where to stow them, large, readable instruments and handy controls. Materials are nice but not extravagant, and if the whole effect is one of overstyling, at least the layout is logical and comfortable. An adjustable steering column is optional, and seatbacks can be adjusted over a small range with a screwdriver and wrench. New seatback locks keep them upright until they're released for access to the adequate, but awkward, luggage area behind.

Disc brakes on all four wheels are powerful, smooth and

consistent. Constant contact of the pads with the discs permits a very small travel of the caliper cylinders and thus allows a large mechanical advantage in the actuating circuit: because of this the Corvette is the only production car over 3000 lb with four discs that can get by without power brake assist. And though our test car's power assist gave adequate feel, we consider this item to be unnecessary. Fade characteristics of these Delco-Moraine brakes are excellent, but their ultimate stopping capability is limited to 24 ft/sec/sec or 0.75-g by tire adhesion. The front brakes locked consistently and evenly at this deceleration rate. Squeal prevails when the discs are cool. Parking brake actuation is by small drum units built into the rear discs; in our experience this arrangement works well, but poor adjustment on the test car limited its holding power to about 10% grades.

One unwelcome change from earlier Sting Rays is the amount of engine fan noise in the current model. Apparently nothing has been changed but the shape of the fan shroud, but the fan even with its viscous drive is a veritable windmill all the time. Perhaps the change has achieved the greater air flow ⟶

1967 CORVETTE
AT A GLANCE

Price as tested...............................$4824
Engine..................V-8, ohv, 5356 cc, 300 bhp
Curb weight, lb............................3160
Top speed, mph...............................121
Acceleration, 0–¼ mi, sec....................16.0
Average fuel consumption, mpg................15.4
Summary: Performance & handling with comfort, reliability & longevity. Rather large & heavy for 2-only seating because of separate body/frame. Good value for money. Probably last year for this body style.

New for 1967 are four-lamp flasher control on steering column, center handbrake, console placement of fresh-air vent controls.

1967 CORVETTE

needed with the air injection smog pump to provide adequate cooling under idling conditions.

All things considered, the Sting Ray is a big value for the money. It matches any of its European competition for useful performance and walks away from most of them; it's quiet, luxurious and comfortable under ordinary conditions; easy to tune and maintain; and even easy on fuel if its performance isn't indulged too often. Quality of assembly is lacking, however, and the following items were amiss on our test car: several rattles; improper clutch adjustment; an air leak over the windshield; choke setting; sticky throttle linkage; and a fresh-air vent that wouldn't shut off.

The improvements we would most wish for in the next Corvette series would be lighter weight, improved body structure and quality control, and a better ride on poor surfaces. But in the meantime the Corvette ranks with the best sports/GT cars the world has to offer, regardless of price. It is significant to note that with the power assists omitted, the Corvette tested can be bought (through the normal haggling process) for approximately $4000.

Emission-reducing air pump is on right front of engine. Wire isn't standard—accordion hose was an emergency replacement.

The handsome new 6-in.-wide standard wheel. Power steering plumbing can be seen inboard, just under frame rail.

ROAD TEST
1967 CORVETTE

SCALE: 10" DIVISIONS

PRICE
Basic list................$4228
As tested...............$4824

ENGINE
Type..................ohv V-8
Bore x stroke, mm.....102 x 82.6
  Equivalent in......4.00 x 3.25
Displacement, cc/cu in...5356/327
Compression ratio.........10.25:1
Bhp @ rpm..........300 @ 5000
  Equivalent mph..........111
Torque @ rpm, lb-ft..360 @ 3400
  Equivalent mph............78
Carburetion........one Holley 4-V
Type fuel required........premium

DRIVE TRAIN
Clutch diameter, in..........11.0
Gear ratios: 4th (1.00).....3.36:1
  3rd (1.46)................4.90:1
  2nd (1.88)................6.32:1
  1st (2.52)................8.47:1
Synchromesh.............on all 4
Final drive ratio.........3.36:1
  Optional ratios..........3.08:1

CHASSIS & BODY
Body/frame....steel ladder frame, separate fiberglass body.
Brake type: vented discs, 11.75-in. diameter; single calipers.
  Swept area, sq in..........461
Wheel type & size....disc, 15x6JK
Tires.....UniRoyal Laredo 7.75-15
Steering type....recirculating ball
  Overall ratio..............17.6:1
  Turns, lock-to-lock..........2.9
  Turning circle, ft............41.6
Front suspension: independent with unequal-length A-arms, coil springs, tube shocks, anti-roll bar.
Rear suspension: independent with lateral leaf spring, lateral struts, U-jointed halfshafts, trailing arms, tube shocks.

OPTIONAL EQUIPMENT
Included in "as tested" price: AM/FM radio, 4-speed gearbox, limited-slip diff, power steering & brakes, exhaust emission control, minor items.
Other: various engine options, automatic transmission, air conditioning, power windows, competition equipment.

ACCOMMODATION
Seating capacity, persons........2
Seat width.............2 x 19.0
Head room...............39.0
Seat back adjustment, deg........2
Driver comfort rating (scale of 100):
  Driver 69 in. tall............90
  Driver 72 in. tall............85
  Driver 75 in. tall............75

INSTRUMENTATION
Instruments: 7000-rpm tachometer, 160-mph speedometer, fuel level, ammeter, oil pressure, water temperature.
Warning lights: parking brake, brake fluid loss, headlights retracted, high beams, directional signals.

MAINTENANCE
Crankcase capacity, qt........5.0
  Change interval, mi.......6000
Filter change interval, mi.....6000
Chassis lube interval, mi.....6000
Tire pressures, psi..........24/24

MISCELLANEOUS
Body styles available: convertible with hard and/or soft top, coupe.
Warranty period, mo/mi.24/24,000

GENERAL
Curb weight, lb............3160
Test weight..............3540
Weight distribution (with driver), front/rear, %....49/51
Wheelbase, in..............98.0
Track, front/rear......57.6/58.3
Overall length............175.1
  Width.................69.6
  Height................49.8
Frontal area, sq ft..........19.2
Ground clearance, in.........5.0
Overhang, front/rear....32.0/45.1
Usable trunk space, cu ft......8.0
Fuel tank capacity, gal......18.5

CALCULATED DATA
Lb/hp (test wt)............11.8
Mph/1000 rpm (4th gear)....23.1
Engine revs/mi (60 mph)....2600
Piston travel, ft/mi........1405
Rpm @ 2500 ft/min........4610
  Equivalent mph...........103
Cu ft/ton mi..............139
R&T wear index..........36.6
Brake swept area, sq in/ton...261

ROAD TEST RESULTS

ACCELERATION
Time to distance, sec:
  0–100 ft..................3.5
  0–250 ft..................6.1
  0–500 ft..................9.0
  0–750 ft.................11.3
  0–1000 ft................13.4
  0–1320 ft (¼ mi)........16.0
Speed at end of ¼ mi, mph...86.5
Time to speed, sec:
  0–30 mph.................3.4
  0–40 mph.................4.6
  0–50 mph.................5.9
  0–60 mph.................7.8
  0–70 mph................10.0
  0–80 mph................12.9
  0–100 mph...............23.1
Passing exposure time, sec:
  To pass car going 50 mph...4.9

FUEL CONSUMPTION
Normal driving, mpg.......14–18
Cruising range, mi........260–325

SPEEDS IN GEARS
4th gear (5500 rpm), mph.....121
3rd (5500)..................85
2nd (5500)..................68
1st (5500)..................50

BRAKES
Panic stop from 80 mph:
  Deceleration, % g...........75
  Control..............excellent
Fade test: percent of increase in pedal effort required to maintain 50%-g deceleration rate in six stops from 60 mph.........16
Parking: hold 30% grade......no
Overall brake rating.....very good

SPEEDOMETER ERROR
30 mph indicated......actual 29.0
40 mph.................39.4
60 mph.................59.3
80 mph.................78.1
100 mph................95.6
Odometer, 10.0 mi.....actual 9.8

ACCELERATION & COASTING

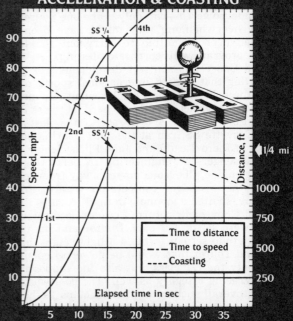

GRAND SPORT

BY BROCK YATES

Long ago in th[e]
General Motor[s]

David Erwin of Painted Post, New York, describes himself as a "Chevy man." There are "Ford men" and "Plymouth men" and "Pontiac men," but Erwin and the close-knit group of friends who gather nightly to labor over one of the rarest of all Chevrolets remain loyal to that marque.

David Erwin could also be cast as a "Ferrari man." A tall, reserved banker, and the heir of a prominent family of landed gentry whose holdings date back to the Revolution, Erwin is the sort of precise, well-bred young man whose fascination with the Maranello product is practically a foregone conclusion. In fact, one stall in Erwin's comfortable workshop behind the family's large, colonnaded homestead is occupied by one of the last GTO lightweights ever built.

But the machine stabled beside the Ferrari is the one that keeps "Chevy man" Erwin and his friends turned on. This is one of the fabled "Grand Sport" race cars, a four-wheeled passenger pidgeon whose existence bears testimony to those mystery-shrouded days when Chevrolet was committed to winning every major sports car race in the world, including Le Mans.

Ever since the spring of 1963, when General Motors summarily cancelled all competition activities, the Corvette Grand Sport has established itself as one of the most fascinating enigmas in motor racing annals. Periodically, one of the five examples Chevy built has appeared at places like Sebring or Nassau, has raced around at shocking speeds, and then has plunged back into mysterious seclusion. People are still talking about how the Jim Hall/Roger Penske Grand Sport stunned the Fords and Ferraris by leading the early laps of the 1964 Sebring 12-Hour, and how Penske thrashed the late Ken Miles and his Cobra 427 prototype at Nassau later the same year.

These rare outings, though impressive, never did the machines full justice, because GM's anti-racing policy was so effective than never once did a Grand Sport reach the race track at the level of readiness that Chevrolet engineers had intended. That the cars were so impressive when they were finally raced—both outdated and underpowered—can only make one pause to wonder how overwhelming they might have been if Chevrolet had been permitted to carry out the full Grand Sport program with corporate blessing.

The Grand Sport was a direct development of the 1958 FIA ruling that limited international sports/racing cars to three liters engine displacement. Up until that moment, Chevrolet had been hard at work on some exotic big-engined sports/racers, the last of which, the Corvette SS, ran at the 1957 Sebring race in the hands of Piero Taruffi and John Fitch. Seeing no benefit to passenger-car engineering in perfecting a 3-liter racing engine, Chevrolet stayed away from road racing until their resident competition wizard, Zora Arkus-Duntov, spotted a loophole in the FIA rules that would permit Chevrolet's return. Because no displacement limits were set on GT cars, Duntov and his talented design group set out to build a lightweight, big-engined Grand Tourer that would be powerful enough to win not only the GT class, but also the supposedly faster sports/racing category as well. The goal was no less than an overall victory at the Le Mans 24-hour classic.

Chevrolet, still shying away from an all-out racing

PHOTOGRAPHY: ALFRED FISHER

CORVETTE GRAND SPORT

car disguised as a GT car, wanted their Grand Sport to look like a production car—in this case, their brand new Sting Ray. Duntov and his Corvette engineers reckoned that it would take 600 horsepower to push the Sting Ray up the straight at Le Mans at a competitive top speed, 4-wheel disc brakes to slow it down, and a vehicle weight of 1800 lbs. to achieve competitive lap times.

Because the rules allowed a GT engine to displace less than its stated capacity, but not more, Chevrolet specified a 327 cu. in. aluminum block with the standard 4-inch bore and a ¾-inch stroker crank (4″ x 4″, in hot rod parlance), for an enormous 402 cu. in. displacement. The heads, also aluminum, have never been seen by the public. The combustion chambers were hemispherical and featured twin ignition. Sitting astride the engine was a complex fuel injection system, with eight long ram tubes poking up through the hood.

Initial tests proved the 4″ x 4″ setup unsatisfactory and subsequent engines were given 3.75-in. strokes, which reduced displacement to 377 cu. in.

The chassis was tubular aluminum with the fully independent Sting Ray rear suspension. The body was slightly smaller than stock Sting Ray to improve the

CORVETTE GRAND SPORT SPECIFICATIONS*

Manufacturer: Chevrolet Motor Division
General Motors Corporation
Detroit, Michigan

Vehicle type: Front-engine, rear wheel drive, 2-passenger GT racing car, ladder frame, aluminum structural members integrated with reinforced plastic body, plexiglass side and rear windows

ENGINE:
Type: Water-cooled V-8, aluminum block and heads, 5 main bearings, twin ignition, hemispherical combustion chambers, two valves per cylinder
Bore x stroke................4.00 x 3.75 inches, 101.7 x 95.3 mm
Displacement............................377 cu inches, 6172 cc
Compression ratio....11.0 to one (alternative ratios: 10.0, 12.0 to one)
Carburetion...................................Port-type fuel injection
Valve gear........Push-rod operated overhead valves, mechanical lifters
Valve timing: Intake opens.............................55° BTDC
Intake closes.............................117° ABDC
Exhaust opens...........................106° BBDC
Exhaust closes...........................44° ATDC
Valve lift..................................0.46 inches (intake and exhaust)
Valve diameter: Intake 2.20 inches
Exhaust 1.72 inches
Power (SAE)..................................550 bhp @ 6400 rpm**
Torque (SAE)..................................500 lbs/ft @ 5200 rpm**
Specific power output.................1.46 bhp/cu in, 89.3 bhp/liter
Maximum recommended engine speed................6500 rpm**

DRIVE TRAIN
Transmission....................4-speed manual, all synchromesh
Clutch diameter...10.0 inches
Final drive ratio......3.55 (alternative ratios: 3.08, 3.36, 3.70, 4.11, 4.56)

Gear	Ratio	mph/1000 rpm	Max. speed (3.55 ratio axle)
I	2.21	10.6	69 mph @ 6500 rpm
II	1.64	14.2	92 mph @ 6500 rpm
III	1.27	18.4	119 mph @ 6500 rpm
IV	1.00	23.3	152 mph @ 6500 rpm

DIMENSIONS AND CAPACITIES
Wheelbase..98.0 in
Track..............................F: 56.8 in, R: 57.8 in
Length..172.8 in
Width..69.6 in
Height..51.9 in
Ground clearance..4.3 in
Curb weight..1908 lbs
Lbs/hp..2.9
Battery capacity.............................12 volts, 61 amp/hr
Generator capacity..624 watts
Fuel capacity..36 gal
Oil capacity...9 qts
Water capacity...16.5 qts

SUSPENSION
F: Ind., fabricated tubular unequal-length wishbones, combined coil spring/shock absorber units, anti-sway bar
R: Ind., lower transverse leaf spring, half-shafts acting as upper locating links, trailing arms, tubular hydraulic shock absorbers

STEERING
Type..Recirculating ball
Turns lock-to-lock.......................................3.25
Turning circle...40 ft

BRAKES
F: 11.47-in Girling discs with 3.30 x 2.04-in pads (power assisted)
R: 11.50-in Girling discs with 3.06 x 1.55-in pads (power assisted)
Swept area.....................................435.2 sq inches

WHEELS AND TIRES
Wheel size and type..6.0 x 15-in, magnesium knock-off type, weighing 16.5 lbs each
Tire make, size and type........Firestone SS170 TW, F: 7.10/7.60-15
R: up to 8.00/8.20-15

*From FIA Application 837 A-63
**Our estimate

> The Grand Sport's engine was to be a 600-hp, 402 cu. in., all-aluminum V-8 with twin ignition and hemi heads.

aerodynamics, although extra-wide fender valances later had to be fitted to accommodate ever-wider tires. Aside from the front headlights being placed behind streamlined plexiglass housings, the cars looked quite similar to the stock Sting Rays.

Chevrolet submitted papers for FIA recognition of the Grand Sport early in 1963, promising to have built 100 examples between July 7, 1962 and June 1, 1963 (Le Mans was on June 15-16 in 1963), but the papers were hastily withdrawn when the Corporation pulled the plug on the racing program. An unauthorized copy of the papers, hoarded for years, turned up in New York recently, and makes interesting reading indeed (see the specifications table, opposite).

As far as we can determine, five coupes were built by Chevy before the racing ban. Chevrolet originally intended to market several hundred production Grand Sports to the public at a price under $10,000, but that aspect of the project never got off the ground, and the "Grand Sport" label was applied only to the factory-built race cars.

Duntov and his crew went to Sebring, Florida, in mid-January '63, for pre-race practice, with Masten Gregory, Dr. Dick Thompson, and Duntov himself do-

GRAND SPORT!

ing the driving. The 377 cu. in. engines were not ready and fuel-injected 327s were used in their stead. It was discovered that the Girling solid discs would last only a few hundred miles, and Chevrolet switched to vented discs of their own design with Girling calipers (presaging the vented discs of the '65 Sting Ray). The test program was judged an overall success, and the Chevrolet people planned to debut the cars at Sebring in March, 1963, followed by an all-out assault on Le Mans that June.

While details for the trip to the Sarthe circuit were being worked out, Mickey Thompson built some lightweight Sting Rays of his own, powered by the top-secret "Mk. II" 427 cu. in. porcupine-head stock car engine, and entered them in the Daytona Continental. In later years, some of the Thompson lightweights were to become confused with the factory-built Grand Sports, but they were quite unrelated to the exquisite machines that Duntov's group was completing in the winter of 1963.

Immediately following the Daytona 500, General Motors lowered the boom. Three Mk. II-engined Chevy stockers driven by Junior Johnson, Johnny Rutherford and G. C. Spencer were so much faster than everything else that only incomplete "debugging" prevented them from turning the race into a private donnybrook. Success seemed a heartbeat away. When GM brasshats Donner and Gordon suddenly announced that the Corporation would henceforth follow the 1957 Automobile Manufacturers Association anti-racing resolution to the letter, the enthusiasts at Chevrolet were probably as shocked as anybody else. The Grand Sport—and with it Sebring and Le Mans—became as dead an issue to GM as the Missouri Compromise.

The cars sat around in some dark corner of Chevrolet Engineering for 10 months. Then the lid opened a crack before slamming shut for another year. In December of '63, three of the Grand Sports showed up at Nassau. Two were entered by John Mecom, while the other was in Jim Hall's stable. The cars were not seen again until Nassau '64, when Penske scored the first of three wins that week, immediately prior to his retirement.

By now the Grand Sports were getting old. Chevy seemed to have forgotten about them. No eyebrows were raised when two of them showed up at Sebring in March '65. No eyebrows, that is, except Dave Erwin's.

From the moment he clapped eyes on a Grand Sport, Erwin wanted one. He was spellbound at the sight and sound of the wickedly powerful machine. He followed the course of the car at Sebring and Elkhart Lake, after which it came into the hands of Pennsylvanian George Wintersteen. Wintersteen, a friend of Penske's, raced the car sporadically —at Sebring in '66, and in some club events. Last fall Wintersteen casually offered it for sale minus engine, and Erwin snapped it up.

The car was in excellent condition. The plastic headlight covers had been replaced with wire mesh; the differential oil cooler, which had originally been attached to the rear deck lid, was missing; the filler cap had been altered slightly; and the air jacks had been removed. Otherwise the car was practically the same as the day it first rolled out of the General Motors Tech Center nearly five years before.

When we were given the opportunity to drive Erwin's Grand Sport (designated Chassis #5) the car was fitted with a muscled-up 327 that Erwin, his friends, and his brother Tom, had stuffed aboard. The engine—which utilizes a Crower cam and hi-rev kit, General Motors exhaust headers and hi-rise aluminum manifold, cast iron heads from the fuel-injection Corvette engine, and a big Holley carburetor—is intended as an interim powerplant. Last May, Dave saw another ad. This one was placed by NASCAR short-track star Bobby Allison, and was offering one complete Mk. II engine, a pair of spare blocks and enough extra Mk. II spares to build another complete engine. Erwin grabbed them, and is hoarding them as if they were H. L. Hunt's oil leases.

Because the blue and white brute wasn't registered (Erwin plans to run the car in a few regional races, then restore it, but never drive it on the street), we towed it a short distance to a long stretch of completed but not yet opened four-lane highway for some test runs. The lads had ingeniously stuffed the mufflers designed for the new Sting Ray external side-mounted exhaust pipes into the enormous Grand Sport tubes, which reduced the noise level to a point where the off-duty personnel in the State Police barracks a few miles down the road wouldn't be unduly disturbed.

It was well and truly a racer. Thumping down the highway on the tremendous Firestone Indy tires, the familiar odors of oil and hot paint wafted into the cockpit, along with the sound of air rushing around the hand-operated plexiglass windows. This mingled with the whine of the fully-locked differential gears and the slick prototype Muncie gearbox.

The gearbox and brakes were nearly perfect. That means stops like the car had just run into a mud bank, while the transmission was as loose—and yet precise—as any we've ever handled. The locked rear end made it an awful chore to negotiate corners under 30 mph, mainly because the inside rear wheel would moan and scuff the pavement, and the rear end sounded as if it was going to explode through its cast aluminum housing, but at high speeds the car was a dream. It had virtually neutral steering characteristics, and we could find nothing in its entire handling range that could be described as treacherous or unstable.

Naturally, "Chevy man" Erwin is considering the installation of one of his Mk. II 427s. Although he admits this would not be an entirely authentic switch, he is correct when he says that the assembly of a hemihead, twin-ignition, port-injected, all-aluminum Chevy engine is out of the question, and he hopes that this will get him off the hook with the historical purists.

In any case, David Erwin of Painted Post has himself one of the most unique automobiles in the world. The other four Grand Sports have been sold and resold several times, although either Wintersteen or Penske has owned all of them at one time or another. Wintersteen and Mecom currently own Grand Sports that have been converted into roadsters, while Texan Delmo Johnson and Toledo, Ohio, Chevy dealer Jim White own the remaining two coupes. So Erwin has one fifth of the total supply of Grand Sports, making his possession all the more valuable.

There it sits, in the loving care of Erwin and his buddies, looking ready to tangle with any GT car built in the half-decade since the car was made. It surely is one of the fastest relics in the world—a car that remains only a few seconds off the best lap times of the fastest road racing cars in the world, even today.

Given an even break, old Chassis #5 might have brought the Corporation that tossed it out like an illegitimate son America's first victory at Le Mans—and three years ahead of time at that. But that's all over now . . . except for "Chevy man" David Erwin.

C/D

CARS ROAD TEST

The 435-hp Corvette is super-boss, super-quick, super-expensive, super-impossible to insure and super-heistable! The big question is

WHAT PRICE GLORY?

BY MARTYN L. SCHORR and JOEL ROSEN

THE BIG QUESTION in performance circles ever since General Motors dropped the ax on divisional race car and parts building activities has been, "When will Chevy come back into racing?" Hip street and strip-racing buffs, however, stopped asking this question many moons ago. In fact, the "in crowd" really doesn't care if Chevy ever comes back into racing, as long as the factory keeps on turning out big-block *imagemobiles*. The truth of the matter is that Chevy will probably never go back to race car sponsorship deals, unless, of course, its fantastic youth-performance image drops off. The big boys at "Number One" realized a long time ago that you don't have to lose money in the racing game to sell street machines!

A prime example of Chevy's marketing genius is the latest and biggest of the big-block Corvettes, the 435-hp "acceleration machine." Here's an honest-to-goodness factory production car that in stock condition is capable of annihilating any street supercar, and running an honest 150 mph (traffic permitting, of course). It boasts three carburetors, 427 "porcupine needles," full instrumentation, four-wheel discs, a sophisticated compromise suspension and an unreal power-to-weight ratio.

How does Chevy get away with violating two of Generous Motors' golden rules (Thou shalt not sell cars with multiple-carbureted engines, Thou shalt not build a car with less than 10-pounds-per-rated-horsepower)? They beat the rap by calling the Corvette a sports car and not a normal passenger car. And since its competition (Jaguar, Aston Martin, Ferrari, etc.) in the GT category is made up of multi-carbed complex monsters, Chevy was permitted to go the route. How many tri-powered Sting Rays will actually end up in the hands of sports car buffs or racers is anybody's guess. We are of the opinion that none will, and that all built will end up on the drag strip or at the Drive-In!

Because of Chevy's rather conservative horsepower rating method we realized long before production time that this Sting Ray would be the dynamite street package of the year, bar none. We knew what well-tuned 425-hp '66 models were capable of, and figured that the '67 tri-power had to be even more explosive.

Getting a factory test car proved to be a little more difficult than we had anticipated. Chevrolet's public relations man, who specializes in smiling, shaking hands and letting you know that they do pretty well without performance publicity, saw no reason to go out of his way and supply us with a test vehicle. After a number of phone calls which netted us with a huge supply of evasive answers and a large bill, we gave up trying. I guess when you're on top you just don't have to try harder!

A local phone call to Joel Rosen at Motion Performance, Inc. (598 Sunrise Highway, Baldwin, New York) proved to be far more effective than the long distance attempts. Joel secured for us a well broken in '67 Corvette ragtop with four speed, 435-hp engine work. Joel also volunteered to assist us with the road test, since he has owned and competed (drag and sports cars) with a great

Minor trim changes highlight the latest and hottest Sting Ray. Louvers, scoop, status plaques and backup blinker assembly are '67 goodies. Note the new wheels.

Co-tester Rosen indicates where hood should be cut to take advantage of the non-functional scoop. Tri-power setup uses Holley pots with 1000 CFM rating.

1967 STING RAY POWER TEAMS

Engine Bore @ Stroke	Horsepower @ Torque at RPM	Carburetion @ Induction System	Comp. Ratio	Cam @ Lifters	Trans-missions	Axle Ratios Standard	Axle Ratios Positraction
STANDARD ENGINE							
327-cu.-in. V8	300 @ 5000	4-Barrel	10.0:1	General Performance	3-Speed (2.54:1 Low)	3.36:1	3.08:1 * 3.36:1
4.00 x 3.25 ins.	360 @ 3400	High-Flow Air Cleaner		Hydraulic	4-Speed (2.52:1 Low)		
					Powerglide	3.36:1	3.36:1
EXTRA-COST OPTIONAL ENGINES							
327-cu.-in. V8	350 @ 5800	4-Barrel	11.0:1	High Performance	4-Speed (2.52:1 Low)	3.36:1	3.36:1 3.55:1 *
4.00 x 3.25 ins.	360 @ 3600	High-Flow Air Cleaner		Hydraulic	4-Speed (2.20:1 Low)	3.70:1	3.70:1 4.11:1 *
427-cu.-in. V8	390 @ 5400	4-Barrel	10.25:1	High Performance	4-Speed (2.52:1 Low)	3.08:1 *	3.36:1
4.251 x 3.76 ins.	460 @ 3600	High-Flow Air Cleaner		Hydraulic	4-Speed (2.20:1 Low)	3.36:1 *	3.08:1 * 3.55:1 * 3.70:1 *
					Powerglide		
427-cu.-in. V8	400 @ 5400	Triple 2-Barrel	10.25:1	High Performance	4-Speed (2.52:1 Low)	3.08:1 *	3.36:1
4.251 x 3.76 ins.	460 @ 3600	High-Flow Air Cleaner		Hydraulic	4-Speed (2.20:1 Low)	3.36:1 *	3.08:1 * 3.55:1 * 3.70:1 *
					Powerglide		
427-cu.-in. V8	435 @ 5800	Triple 2-Barrel	11.0:1	Special Performance	4-Speed (2.20:1 Low)	3.55:1 *	3.36:1 * 3.70:1 * 4.11:1 *
4.251 x 3.76 ins.	460 @ 4000	High-Flow Air Cleaner		Mechanical			

*Available as Positraction only

Top-line 435-hp mill utilizes last year's big-valve heads, forged pistons, solid-lifter cam and good rods. Tri-power is new addition.

variety of Sting Rays since the first was introduced in '63. The car's owner, Mike Zarinski of Bayside, New York, had just returned from a cross-country trip and was more than happy to loan us his "guided missile" for a test. Since the car was privately owned and not a factory car, we limited our testing to a couple of days and did away with some of our *banzai* blast-offs.

What's it like driving a 435-hp Corvette? It's sort of like guiding a four-wheeled, two-passenger rocket sled with license plates and an AM-FM radio and stereo tape system! It gets you wherever you want to go quicker and faster than most anything on the road. In the hands of an inexperienced or faint-of-heart driver it's a lethal weapon. All this and our test car was only a wee bit faster from the top of third gear up than the average 425-hp '66 model!

Finished off in rather well-applied deep green paint with a unique white race strip applique on its scooped hood, our test car had already logged 12,000 miles on its odometer. Owner Zarinski had had mucho trouble with the engine while making his cross-country jaunt and had to rely on dealer facilities for service work. The car was plagued with excessive oil consumption since new and a local Arizona dealer solved the problem by installing a couple of over-size valves and new seals in one head. Zarinski also managed to float the valves while in New York and ended up at Motion Performance for some new springs and pushrods.

On the road the car was unusually sluggish (for a car of this type) through the first two gears, and we felt that a good-running 425-hp model would have no problem doing a job on our test car. Once we reached the top of third gear the engine came on strong and left us with the impression that it was strictly a top end machine. The stock 3.55-to-1 gearing would allow the car to go an honest 150 mph (tires and road conditions permitting), but was actually preventing the car from "honking" on the bottom. Zarinski chose this *standard* gearing because many cross country trips are on the plans board.

With the stock high-performance tires mounted on the new 6-inch-wide wheels it was difficult getting a bite once the revs were brought up for a maximum performance burn-out. Considering the tires and the gearing, we did most of our acceleration tests by coming out easy, then nailing it. Even though the car felt weak on the bottom, it was powerful enough to plaster driver and passenger to the backs of the buckets through three gears.

We also felt that the car's performance potential was hindered slightly by the four-speed linkage supplied by the factory. In the past, most Corvettes equipped with four-speeds had the nasty habit of transmitting some vibration to the pilot via the shift linkage. Well, all traces of shift line vibration are gone, but the linkage itself is too stiff for quick power shifting. It's too easy to get hung up in neutral and "spill the beans." The new Muncie linkage setup runs off a chassis-mounted bracket which also supports the pulleys for the cable-controlled rear parking brake.

Owner Zarinski, right, explains the operation of his super-efficient burglar alarm system to Rosen. This type of unit is a must for 'Vettes in the New York area!

Instrument layout and new console-mounted parking brake lever are plus-features. Tach and shift proved to be useless for performance driving and power shifting.

After the test, Zarinski went for a series of engine modifications along with a recommended Hurst Competition-Plus linkage conversion.

Under the scooped (non-functional) hood of the biggest of the big-block Corvettes rests a 427-incher ($437.10 extra) that is similar to last year's 425-hp model except for tri-pot intake plumbing. The Chevy boys have tried every possible carburetion combination on the Corvette since its introduction in 1953. They started off with sidedraft tri-power on the original Six, went to single four-barrel and dual inline quads on the 265-283 Eights, advanced to fuel injection and single tuned quads on the 283-327, reverted to single quad carburetion on the 396-427 big blocks, then changed over to three monster Holleys on the top-line '67 427.

The tri-power engine which goes for one-dollar-per-horsepower, carries 435-hp at 5800 rpm, 460 foot/pounds of torque at 4000 rpm ratings and is graced with such goodies as 11-to-1 forged pistons, a solid lifter .519-inch lift camshaft, transistorized (who needs it?) ignition and well-designed, tuned cast-iron exhaust manifolds. The all-new tri-power package includes three Holley two-throats, a cast aluminum manifold and a triangularly-shaped air cleaner assembly.

Performance characteristics of the three-two setup are unlike those usually associated with a tri-power vehicle. A newly-developed secondary carburetor control system eliminated the abrupt throttle opening characteristics usually found in most passenger car setups. Just for the record, the driver controls the operation of the center carburetor (good low end performance and economy, smooth idle speed), while venturi vacuum of the center carb controls the opening of the outer carbs. Mass air flow through the center carb dictates end carb opening. The Corvette's design utilizes vacuum sensed in the carburetor venturi rather than elsewhere in the engine, resulting in more accurate mixture feeding.

During full-throttle acceleration we discovered that the secondary throttle blades started to open at approximately 2000 rpm, and opened fully at 4000 rpm. From 4000 rpm up all six throttle blades stayed in the full open position. The flow rating of the three Holleys is 1000 CFM as compared with 850 CFM for the *optional* big Holley single four-barrel which was considered *the* "hot setup" for this engine.

Even with the weight and the power of the 427 up front, the Sting Ray is still a super-handling automobile. It's road manners may not be as precise as some of the *Continued on page 116*

1967 CORVETTE STING RAY SPECIFICATIONS

ENGINE

Type	OHV V-8
Displacement	427 cubic inches
Compression Ratio	11.0-to-1
Carburetion	Holley Tri-power
Camshaft	Solid Lifter, .419-inch lift
Horsepower	435 @ 5800 rpm
Torque	460 foot/pounds @ 4000 rpm
Exhaust	Dual headers, dual pipes
Ignition	Transistorized

TRANSMISSION

Make	Muncie four speed
Control	Inland Steel floor shift
Ratio	Close 2.20-to-1 first

REAR END

Type	Positraction
Ratio	3.55-to-1

BRAKES

Front	11.75-inch power-assisted discs
Rear	11.75-inch power-assisted discs

SUSPENSION

Front	Independent, HD coil springs, HD shocks, sway bar
Rear	Multi-leaf, torque control arms, independent
Steering	Manual
Overall Ratio	20-to-1

GENERAL

List Price	$4141
Price As Tested	$5300
Weight	3400 pounds
Wheelbase	98 inches
Overall Length	175 inches
Tire Size	7.75 x 15 "Redline" Goodyears

PERFORMANCE

0 to 30 mph	2.5 seconds
0 to 60 mph	5.0 seconds
Standing ¼ mile mph	111 mph
Elapsed Time	12.90 seconds
Top Speed	150 mph (est)
Economy	12.0 mpg

higher priced "sewing machines" which are pawned off as GT machines, but there's nothing else on the market that comes close to the Sting Ray's performance and price tag. On good roads, the relatively stiff independent front and rear suspensions do a good job of dampening out irregularities. On rough roads, the driver and passenger are forced to chip in and help in the dampening, however. Handling control is positive, body lean on corners is non-existent and steering is as neutral as possible with the front-mounted 435 hp load. Because of the power on tap it's almost possible to do away with the steering wheel under some conditions and steer via throttle control. In the hands of an experienced driver the Corvette can be made to do the impossible. A slight improvement over '66 was noticed because of the extra ½-inch width wheels used all around in '67. As usual, the four-wheel discs with power assist were firm, positive and are capable of safely stopping a car of the 427 Sting Ray's caliber.

Inside the '67 there are a couple of changes worth talking about. The parking brake control is now to the rear of the shift assembly and is of the race car lever design. Cables run from the stick via pulleys mounted off the shift linkage bracket to special drums incorporated in the rear disc design. The fly-off handle is neat and convenient, but it does away with the normal third party ating position. This is good from an insurance company's point of view, but it further limits the already-limited passenger space. Third passengers, especially females, will have to think a few times before accepting a ride in a '67 Ray!

We were not impressed with the vinyl interior (leather optional at extra cost) and felt that anyone spending over *five thousand* greenies deserves a little more in the line of quality upholstery. As always, the Corvette's instrumentation is super-convenient and complete, except for the lack of a good tach. The factory unit is well-placed and easy to read, but its mechanical cable drive leaves a lot to be desired when you're running redline rpm with a big-block motor. It's especially bad in the high rpm range where shift accuracy is of the utmost importance.

Thanks to extremely mild styling changes, the '67 version is even more of a head-turner than its predecessors. Identification emblems, new fender louvers and a scooped hood add to the already exotic basic lines of this truly ultimate supercar. Bridging the gap between standard wheels with hubcaps and the ultra-exotic aluminum finned wheels ($245 extra) are drilled-out sports wheels with center caps and beauty rims. From a distance they give the appearance of specially-crafted custom wheels which add even more to the image appeal of the car. The super wheel option this year is a bolt-on finned aluminum wheel which replaces last year's out-of-hand knock-offs.

Although economy has never been the Corvette's forte, we were amazed when we checked over Zarinski's cross-country trip records. Even at legal highway speeds and up he was able to realize almost 15 mpg. Think about that for a while! How many GTO, Street Hemi or other such street supercars can even come close to this mark? Zarinski admitted, however, that around town gas consumption dropped down to approximately 9 mpg. This drop he credited to cleaning out the engine at traffic lights and other assorted fun and games! Thanks to the standard 20-gallon fuel tank it's possible, even with the highest output engine, to do some very grand touring.

Our test car was pretty well representative of the average top performance Sting Ray sold for street use. Gearing, transmission, suspension and interior were all standard items. Optional equipment for this bomb includes, four-series rears, M-22 heavy-duty racing four-speed, F-41 beefed front and rear suspensions, J-56 super-duty disc brakes, NO-3 36.5 gallon fuel tank. The optional off-car parts list makes such unreal equipment as special rods and crank, pistons, valves, aluminum heads, deep pans, baffles, ignition, etc., right up to a complete L-88 super-duty 427 engine available to the big-block drag enthusiast.

We were not completely impressed with the test car's acceleration and felt there was little or no improvement over the 425-hp '66 model. However, after completing our test we had the good fortune to run some competitive tests with a couple of other 435-hp stockers and did discover an obvious improvement over last year's hot one. A good-running 435-hp Corvette is definitely stronger throughout the torque range than a well-tuned 425-hp model. It's also more economical, idles smoother and is impossible to beat when all six barrels are wide open.

Because of the high value of Corvettes and big-block engines at Midnight Auto Sales outlets, owner Zarinski fitted his car with a most efficient theft prevention alarm system. In fact, it's one of the best we've seen to date. Once the car is locked up the alarm will go off if the hood or doors are opened, if the ignition is jumped or if the clutch pedal is depressed. This is sort of a must in New York City if you expect to find your car where you parked it!

Even though the big block Corvette is one of the most desirous cars on the supercar market, Chevy and potential owners are running into big problems. Insurance companies are refusing to insure young and old drivers of Corvettes with larger-than-standard 300-hp small motors. They have classified the Corvette as "a showoff vehicle that gets people in trouble" and as "a natural for car thieves," and they want no part of insuring one. This move forces Corvette owners into state-assigned risk plans which means $800-plus-per year insurance costs. This has caused quite a bit of concern at Chevy where things are really rolling on the 427 bombs.

Now that the Sting Ray is in its fifth and most likely last year of production in its present shape (and name), a lot of buyers are waiting for '68. It's hard to imagine Chevy coming out with anything wilder than its current "rocket sled," but there's a lot of talk going around about a midship-engine Mako Shark. Would you believe . . a 10-second production Corvette? Of course, it won't be insurable and you'll need a competition license before they'll let you on the street with it. But it'll be boss, man. Real boss!

AFTER THE NEW WEARS OFF: CORVETTE STING RAY, 36,000 MILES LATER

BY RON WAKEFIELD

I TOOK DELIVERY of my Corvette Sting Ray convertible on June 19, 1964 from Patterson Chevrolet Co., Birmingham, Mich. I had tried to order a car with minimum extras—radio and 4-speed gearbox only—but was informed that orders were not taken for Corvettes after about June 1 of a given model year because of the close relation of supply and demand. So I shopped a few dealers for both price and the ability to supply me a car equipped as I wanted it. Patterson came closest with a car that had only one additional item, the limited-slip differential, and a very reasonable price. Sticker price came to $4510 and I, dealing as a normal customer (albeit with a little knowledge of the dealer's cost) got the car for $3825. A 4% sales tax and a half-year's licensing brought the total amount to $4100 even.

Competition is stiff for a metropolitan car dealer, and the pressure to cut prices sometimes cuts the dealer's *gross* profit to around $150. Such dealers must sell a high volume of cars, and often the service department cannot devote a proper amount of time to preparing a new car for delivery. However, since the dealer was grossing about $400 on my Corvette, I reasoned that I might expect the car to be well prepared. The car was delivered to me with no more than a wash job, apparently. Problems: a broken spark plug; the side windows wouldn't roll up with the doors shut; splotchy paint on top cover panel; the top didn't fit properly; the passenger's seat rattled furiously; leaks around the windshield; and the steering had a massive squeak in it.

Details of the experience of getting these faults corrected approach the sordid and don't bear going into here; since then I've come to the conclusion that the condition of my car was typical and the dealer service was typical; in my experience it seems most Chevrolet dealers look upon the Corvette as some kind of Funny Furrin Car.

The next shock was insurance. I was at the time working off a few minor traffic violations (Detroit police don't like MGs using their ability to maneuver in traffic), and premiums run rather high on Corvettes to begin with. I had passed age 25, fortunately, and got by for $374 the first year. Later the rate came down.

The rest of the story for the first 36,000 miles is more pleasant. I finally readjusted the convertible top myself; the weatherstripping in the top for sealing the door windows couldn't be adjusted sufficiently to allow the windows to roll up easily, but I must say that the windows do seal well now! The other problems were taken care of by the dealer in his good time.

At 7500 miles the car was pulling slightly to the left, and

I had the front end aligned and wheels balanced for $12.95 before embarking on a cross-country trip. During the trip, at 8209 miles on the odometer, the tach suddenly fell to zero and the engine lost half its power. The tachometer drive gear in the distributor base had jammed and stripped itself, and in the process jerked the distributor around, throwing the timing off. A Louisville, Ky., dealer didn't have the necessary gear but did reset the timing and point gap for $3.50. I completed the trip without a tachometer and later had this repaired on warranty.

The long, hard Michigan winter brought out both the best and worst in the Corvette. I had been driven to buy this rather large, heavy car because the climate forced me to appreciate its powerful heater (a string of British sports cars made me appreciate heaters) and its rustproof plastic body—the salt mines are on overtime up there in the winter. Cold starting had something to do with it too, for none of my British jobs could be coaxed to start at much below zero F. Sure enough, the Corvette started instantly all winter; all that was necessary was one poke of the throttle before turning the switch. And that heater—wow! After a few miles at 15 below the blower could be cut back to its lowest (of three) speed and the temperature control could be reduced from maximum. And it was certainly nice not to worry about those cakes of salt on the bodywork.

That was the best part. The worst was the performance of the Positraction limited-slip differential on glazed-ice roads. This option is just fine for getting one out of the mud or off the line at the drag strip, but when both wheels are in sub-marginal traction conditions, the limited-slip can't make up its mind which should have the torque, and the car proceeds down the road mildly fishtailing from side to side. Under these circumstances, all the little old ladies in their Valconbler II sedans drove right past me and my white knuckles. The original-equipment tires supplied by Chevrolet didn't help things either, but tightwad that I am, I was determined to wear them out before replacing them with something more suitable.

The owner's manual for the Corvette suggests an oil and filter change and chassis lubrication every 6000 miles, and an engine tune-up every 12,000 miles, with the usual admonitions about doing these things more often under adverse conditions. Because most of my driving is such that I warm up the car thoroughly on each trip, I stuck to the maximum intervals on the lubrication and oil changing. I had the first engine tune-up—points, plugs, condenser and carburetor adjustment—done at 13,791 miles by a Chevrolet dealer for $33.85. Not that the engine needed it badly; but I was still proud of the new car and wanted to give it the best.

The winter brought on another problem. Where the clutch linkage enters the bell housing, there is a rubber boot. The boot broke (and has regularly broken since), allowing the salty slush into the clutch where it rusted the throwout bearing solid. The symptom of this condition was that a binding could be felt in the clutch pedal when the engine was running, but not when it was shut off. The dealer had some difficulty in figuring this out—as did I—and replaced the bearing on warranty after finally diagnosing the trouble. In the meantime, I had had the boot replaced at a cost of $3.65. The bearing was replaced at 19,555 miles.

At 20,000 miles a loud "pop," heard on applying or releasing power, developed in the rear end somewhere. Two dealers weren't able to find the cause, and one did replace the bracket which secures the differential to the frame. This didn't solve the problem, and to date I still have it. I now think that it originates as a binding in the limited-slip unit, and hope to report on this later.

It took me less than a year to use up the 24,000-mi warranty, and 24,000 miles were passed with no regrets except for the popping differential. As the engine was quite smooth and peppy, I decided to postpone the specified tune-up for a while.

The original U.S. Royal 800 tires may not have been star performers, but they wore well and evenly and required replacement at 29,500 miles. A new set (four) of 205-15 Pirelli Cinturatos set me back $277, but the newfound security in the rain made them seem well worth it. I selected the 205-15 size because of its approximation to the rolling radius of the original 6.70-15s, but in retrospect I think 185-15s would be better for the car. The left front tire rubs a frame rail when cranked hard over to the left, and steering effort is greater than it would be with the smaller size. Ride harshness, already great in the Corvette, is worsened by these tires and getting them properly balanced has been a problem. Still, all things considered, they are good for the car. I retained the original cross-ply spare, and from experience can say that it's absolutely *verboten* to run any distance on three radials and one cross-ply.

At 30,093 miles it was necessary to straighten the left strut rod in the rear suspension for some unknown reason; the dealer (in California) straightened it because there was no replacement rod in stock. Realignment of both front and rear suspension was also performed at this time, plus wheel balancing all around. Total cost for these operations was $34.20.

At 34,000 miles the throttle pedal became arthritic and I replaced it for $1.90. At 35,000 the cool-side tailpipe rusted through and was replaced for $15.86. It should be noted here that I installed a set of straight-through mufflers, given to me as a gift, a bit earlier; otherwise there might have been a muffler replacement in the repair totals.

Oil consumption was a little heavy at first (800 mpq) and it took some heavy-footed driving to get the rings seated. Since then, the usage rate has been about one quart per 1400 miles. Fuel economy for the relatively mild 250-bhp engine has been agreeable: 18.1 mpg overall, corrected—the odometer reads 102.3 miles for every 100 traveled. The engine does, however, require premium fuel.

Certain observations are in order to explain the rather low operating cost for the period. For instance, at 36,000 miles the engine still didn't need a tune-up; also, because I do most of my own washing, polishing and de-rattling, there's very little money spent for this kind of maintenance. And as the car was not quite two years old when it turned 36,000 miles its depreciation for the mileage is relatively low ($3000 value as of May 1966).

CORVETTE STING RAY

Repairs & Replacements in 36,000 Miles

Change steering to fast ratio, reset toe-in at 915 mi.	$ 9.00
Replace broken spark plug, correct seat rattle, adjust weatherstripping and top at 1816 mi.	warranty
Align front end, balance two wheels at 7500 mi.	12.95
Reset timing and points at 8209 mi (failed tach drive).	3.50
Replace tachometer drive.	warranty
Tune engine at 13,791 mi.	33.85
Replace rubber boot, clutch linkage at 18,469 mi.	3.65
Replace clutch throwout bearing at 19,555 mi.	warranty
Replace differential bracket at 23,629 mi.	warranty
Tires, 4 new at 29,500 mi.	277.00
Straighten rear suspension rod, align both ends, balance all wheels at 30,093 mi.	34.20
Repaint nose, damaged by stone.	6.00
Replace throttle pedal at 34,000 mi.	1.90
Replace tailpipe at 35,000 mi.	15.86
Total repairs and replacements in 36,000 mi.	$397.91

CORVETTE STING RAY

I have not doted on this car, but I have maintained it well. At the end of the 36,000 miles it was performing almost as new. Obviously it is a long-life car, and one requiring relatively little attention. The higher-powered versions of the Corvette require considerably more attention and get much poorer fuel economy, so we're examining the Corvette that's the most economical to run.

The drum brakes of the early Sting Rays are both good and bad. Good, in that at 36,000 miles the linings are barely half-worn. Bad, in that they often pull unevenly, are able to achieve only 62%-g deceleration, and fade rapidly with hard use.

One of the car's weakest features, I have concluded, is one of the features that attracted me to it: its body construction. It is rustproof, to be sure, but it also has a great propensity for rattles, squeaks and general structural shake on rough roads. De-rattling has become part of the Saturday wash-polish ritual. To balance out my evaluation of the body, I must say that the trim, exterior and interior, was fitted properly—and that the Corvette's convertible top is one of the best. It goes up and down very easily, seals well and wears well too. From the looks of it at this point, I'd say it will last the life of the car.

Another strong point is the AM/FM radio. This is the only radio that can be ordered in the Corvette, and it's not cheap. But it has excellent tone quality and reception and has made many a mile of freeway driving tolerable when it would have been otherwise sleep-inducing. A low noise level makes listening possible at speeds up to 80 mph with the top up.

CORVETTE STING RAY

Overall Cost per Mile for 36,000 miles

Delivered price	$4100
Gasoline	660
Oil & "service station" bills	80
Tires	277
Maintenance & repairs	121
Licensing, 19 mo. @ 12.70/year	20
2 mo. @ 56.00/year	9
Insurance	466
Washing & polishing	30
Total expenditure, 36,000 miles	$5763
Retail value at end of period	3000
Cost of driving 36,000 miles	2763
Overall cost per mile	7.68¢

Even with its smallest engine the Corvette can be quite exhilarating to drive; there's always a great reserve of torque and roadholding on tap, so this is no dullard in any sense. It does serve to show that the car can be owned and operated for a very reasonable outlay. Some of the qualities that attracted me to it are superfluous in the mild California climate where I now live, but I suspect I'll put on at least another 36,000 before replacing it.

STING RAY

continued from page 62

stronger version has four-bolt main bearings. The Corvette will get the latter.

The main thing we were concerned with in our tests was evaluating what effect the added weight of the new engine would have on handling. We're talking about some 148 pounds over the standard 327-inch engine. Though the dog-bone track left something to be desired in ascertaining this, we were amazed to find only the slightest difference in handling between it and the 350-horsepower car. The steering was a bit heavier, but there was more than enough extra torque to help aim it. Neither car had any competition options such as tires, shocks, or sway bars, so we feel a fully-prepared 396 should handle at least as well as a fully-prepared 327. In short, it looks good, real good.

Among the options that will make the car competitive is a new four-speed gearbox designed to handle a "prodified" version of the engine. An estimated 470 horsepower is expected with application of allowable SCCA modifications. We didn't have a chance to test the box. We understand it is quite a bit noisier than the standard Muncie. This and a transistorized ignition system are the only other options we legally *know* about.

Meanwhile, the street version of the Mark IV is quite a machine. It goes like the proverbial scalded cat, will be cheaper than an injected 327 model, and has all the attributes of the normal Sting Ray to boot: quietness, weatherproofing, and comfort. The big question, "Will it beat the 427 Cobra?" is a long way from answered yet. We don't know. There are a goodly number of Chevy-loyal competitors who have already placed their orders, so we shouldn't be long finding out.

ROAD TEST: SHELBY GT 500 & 427 STING RAY

A contender for the throne meets the renowned king of speed, and the result is a clash between America's only pair of sports cars.

BY STEVE KELLY

PHOTOS BY RANDY HOLT, JR., GERRY STILES

67' GT handling isn't as good as forerunners', but above average "performance" cars'.

For the first time in its decade and a half of production, Chevrolet's Corvette — billed as "America's only true sports car" — is being challenged for the title. Shelby American's GT 500 Mustang-based sports fastback has all the earmarks of a "purist" type sporty-car with a just-right combination of comfort and convenience features.

Following Shelby American's announcement of a 428-cu.-in. V-8 engined GT, we were skeptical if this car would be suited for street duties. We've driven many small cars with huge engine transplants, and somehow they've always seemed to lack tameness during regular operation — unless regular operation was a full-throttle run down a strip. We found out, though, just how wrong we were after depositing ourselves behind the custom wood steering wheel of an air-conditioned GT 500 with Cruise-O-Matic transmission.

We could hardly believe its smoothness in bumper-to-tail-light traffic and had difficulty realizing it was built by the same people who used to peddle a rough-riding, hard-steering Mustang fastback with Bunyanesque brake pedal and an engine which would outshout a John Deere tractor.

In view of this, we arranged for an even hairier GT 500, and also set up an equally equipped Corvette to see if there really was a challenge.

The basic Shelby GT 500 is a true sophisticate compared to earlier cars built by the famous Texan. There was a brief period when the Ford Cobra, also built by Shelby, was thought to be a fair and equal competitor to the 'Vette, but a too-high price and rather impractical design for a street-destined car put it out of contention. Chevy has cause for some worry from the GT 500 though. There's more room inside, and it's easier to drive in traffic. There's a civilized luggage compartment that's accessible from the outside, and there's at least the same measure of racy styling. The GT 500 runs quieter, and in general is a more practical car for everyday use.

The Corvette can't be sold short though. Even in stock condition, it outruns the GT 500 by a solid 0.7 second in the quarter-mile, stops in less distance from any speed, is easier on gas, and exhibits noticeably better workmanship throughout.

The 'Vette started life back in 1953 — before the world or Ford Motor Co. had ever heard of a car called the Mustang — and was then equipped with an impotent 6-cylinder engine and fiberglass body. The 'glass body has remained, but the 6s were dropped two years later and the car since then has steadily climbed higher and higher in the esteem of performance addicts. Equipping the Corvette with a 427-cu.-in. engine was a natural move for Chevrolet, and for the past two years this version has been the image-maker of the line. Previous editions of the 'glass-bodied car were offered with as much luxury built in as speed and performance, and we remember a statement concerning Corvette sales cutting into

120

Stop-off at Irwindale Drags quickly revealed GT 500's popularity with ¼-mile fans. Most said they preferred it to 'Vette.

'Vette makes tight turns a simple operation; standard tires hinder serious contests.

Cadillac. The luxury is still there, but a hard ride comes with it, so we doubt if Cadillac is still worried.

There's more potential for making the Corvette a hard runner than there is the GT 500. The 428-cu.-in. V-8 in the Shelby car is too heavy for serious work, and it has a rather restricted breathing system compared to the 427-cu.-in. Chevy V-8. Called "porcupine" by Chevy engineers, the valve layout of the 'Vette engine is more ideally suited to getting big charges of fuel/air mixes into the combustion chamber, and the head configuration above the chamber is a "semi-hemi" type, similar to early Chryslers.

Weight distribution is better on the GT 500, with 56.4% being carried on the front wheels. The two total out almost equally, but so much of the Corvette *is* the front, that it is distinctly nose-heavy. This characteristic is abruptly apparent when one *attempts* to leave the line under hard throttle, or when you bring it around a sharp bend while applying power. The rear end becomes frighteningly "light."

Starting with two basic cars, more is standard equipment on the GT 500 than on the 'Vette, but there's some chicanery involved. For the base price of $4195, a GT 500 buyer gets the 428 engine, front-disc/rear-drum brakes, full instrumentation, a 4-speed transmission, and all the modified body work that goes into making a GT 500 the distinctive looking car it is. He also gets the beefed-up undercarriage which includes stiffer springs, adjustable shocks, and modifications to the front suspension to make it corner more flatly. On top of this though, he *must* buy power brakes, power steering, integral roll bar with inertia-reel shoulder harnesses (the best we've ever seen) and a fold-down rear seat. This comes to a total of $264.77 before he ever starts ordering a radio, air conditioning or whatever else he may choose. This is sort of like selling a car without seats and a steering wheel, but putting them on the car at the "buyer's request." All GT 500s are built with this batch of options which are called "standard equipment," but which really are not.

For the base convertible price of $4327.50, a Corvette buyer has all options still in front of him, with nothing mandatory except making the payments. The standard Corvette includes a 300-hp, 327-cu.-in. V-8, either a soft folding top or fiberglass lift-off type, and all instruments and gauges along with a 3-speed all-synchro gearbox. From there the owner can choose any one of four optional engines and two 4-speed manuals or Powerglide transmission. Our price and accessory list on page 29 shows the other options available to outfit the Corvette for street or track duties.

The primary function of an automobile, no matter what type it is, is to carry passengers wherever they may want to go. The GT 500 does this better than the Corvette. It has more room inside for people and packages, and will carry four adults for a short time or

GT 500 vs. STING RAY
continued

two children for a long time without complaints. The fold-down rear seat can become a parcel counter when only two are aboard, and the trunk bulkhead swings up for stowage of skis and the like.

The 500 is much easier to drive in traffic as it's not as low as the Corvette, and it's not as ticklish to keep running at slow speeds. That big 435-hp 'Vette engine likes to work hard and when it's in bumper-to-bumper traffic, it objects.

The ride on city streets is much better in the GT 500. There's very little bumping around, whereas the position of the driver in the Corvette is very close to the rear wheels, and any rebound action from them is strongly noticed by the man at the wheel.

Vision in the two cars is almost equal, but still not excellent. We liked the '66 Shelby GTs for their rear-quarter windows, but construction differences on the '67 Mustang prevented their continuance on Shelby's version. Corvette convertibles have blind rear quarters, too, and the lift-off 'glass top isn't much better. The fastback model allows full vision.

We liked the interior layout of the GT 500, but it doesn't have the Maserati look of the Corvette. The 'Vette has *all* necessary instruments right in front of the driver and all controls within a few inches reach. The Shelby car has the speedometer and tach right in front

Nose and hood of GT are fiberglass items. Scoops deposit air on top of carburetors.

Spoiler is integral part of 'glass deck lid; à la T-Bird tail lamps really light up rear.

Initial test GT had air conditioning and Cruise-O-Matic, making it "sporty-plush."

Second GT lacked optional comforts but ran faster. Shelby adds center two gauges.

"Speedway 350" tires break loose easily, but need to for the best acceleration times.

Clean 'Vette front end is product of "hiding" lights. Hood scoop can be opened.

Standard, horizontal back-up light throws excellent beam, distinguishes '67 Sting Ray.

Standard tires resist high-rpm starts by producing slow times and plenty of smoke.

of the pilot, but the amp and oil pressure gauges (Shelby additions) are positioned centrally below the radio. It's not hard to see them, but they're not as readable as in the Corvette, nor are they in a direct line of sight with the road.

The really impressive points of the GT 500's insides are its great-feeling wood steering wheel and the integral roll bar with inertia-reel shoulder harnesses. The wheel is one of the most comfortable we've ever had our paws on, with a smooth lacquered finish and genuine "sporty" look. Shoulder harnesses can be cumbersome to attach, and restrict the normal movements of the driver, but not so in the GT 500. They fit around you like suspenders. The inertia retractor in the roll bar allows slow movement, but quickly holds you against any sudden jerk or action. You learn to be leisurely when reaching for the cigarette lighter.

The 'Vette spare is in a panel below the rear underside which, in the rain, may give the Triple A reason to upgrade its rates. There is room for perhaps two suitcases behind the seat.

It takes a while to get used to the stiff-riding Corvette, and the low driver position. Once we became oriented, however, we went along with the "true sports car" claim. The 'Vette is a lot of fun, but discretion must be used in driving around town. It's awful easy to bound past speed limits unknowingly.

Less of an all-out sports car, the GT 500 is more at home on the street than on the track. We received many more comments on the 500's styling than we

Corvette gauge layout is nothing short of great, as is close-gate 4-speed shifter.

Entry to neat interior is easier than exit. Hand/parking brake is between seats.

GT 500 vs. STING RAY
continued

did on the Chevy sportster's, but this can be attributed as much to its being new as anything.

Our forecast of how the cars would compare on the track turned out to be accurate. We were highly impressed with the solid-as-a-rock cornering of the GT 500, and its agility in quick turns. It ran impressively in acceleration tests, staying straight when the tires were spun, and it stopped quick and straight. But even with this good performance record, it fell short of outrunning and outmaneuvering the Corvette.

Unfortunately, Chevrolet couldn't find enough room under the fender wells for a tire comparable to the E70-15 nylons standard on the GT 500, so it suffers in this department. Standard 7.75 x 15 rayon tires are fitted, and this makes cornering and starting difficult. Gobs of wheelspin is all that results from a high-rpm start, but proper feathering of the gas pedal will help. The independent rear suspension gets an unfair shake when these regular tires are installed, and the rear end tends to leave the course when the car is pushed. Hooking up tires equivalent to the Speedway 350 Goodyears on the test GT 500 solved the Corvette's problems quickly. The potential is built in, but anyone desiring to put a 'Vette in proper handling form will have to start with a set of tires and the realization that lock-to-lock cornering will be slightly restricted.

It's quite a foe that Shelby's GT 500 has taken on, but not an unconquerable one. The Corvette is designed as a sports car, and that's what it is. It suffers somewhat as a street machine, but in no way is it reminiscent of early English sports cars with fold-down windshields and side curtains.

The 500 has more passenger car than sports car feel, but this could hardly be helped as the Mustang from which it stems has this quality. With a bigger engine (horsepower, not displacement), it would be close to the Corvette's acceleration times.

The two cars are apart yet fairly close. Shelby-American built the GT 500 with the idea of getting more customers for street-type vehicles than they could with the race-oriented GT 350s of the past. Oddly enough, this is what Chevrolet's theory has been too. They've been building a "hot" car for the street that would qualify for competitive use. Shelby's just reversed the game and taken a competition-type car and turned it into a street machine. Both gain in some respects and suffer in others from the compromise. /MT

Big 428 engine fills out GT compartment, leaves little space for working. Progressive linkage is used on dual 4-barrels.

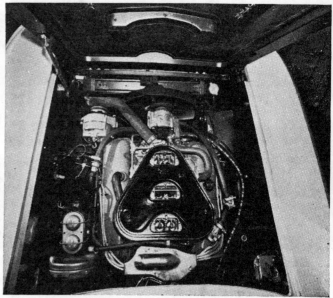
Center carb of 427 has choke and is linked to other two by progressive hookup. Working room around V-8 is very good.

GT's shoulder harnesses are mounted to well-padded and fitted integral roll bar.

Inertia retractors stop sudden motion of GT wearer; allow slow, easy movement.

Hood pins are now wired to GT to prevent loss, or scavenger hunters from scoring.

427 STING RAY: REIGNING KING

- easily opened-up hood scoop
- 435-hp, 427 V-8 with 4-speed
- full instrumentation in front of driver
- center fuel filler
- underside-stored spare
- hide-away headlights
- 4-wheel disc brakes
- independent rear suspension
- forward hinged hood
- luggage compartment accessible from interior

Five-year-old Sting Ray body style still looks good, and all-around quality of fiberglass body is of a very high standard.

SHELBY GT 500: CANDIDATE FOR TOP BILLING

- fiberglass hood, spoiler and scoops
- full instrumentation
- interior vent scoop
- high-speed spoiler
- high-beam road lights
- 355-hp, 428 V-8 with 4-speed
- integral roll bar with shoulder harnesses
- functional rear brake air scoop
- 3-lamp tail lights
- stiffened suspension
- E70-14 Speedway tires

Well executed GT design results from extended nose, rear spoiler and scoops. 'Glass add-ons could stand quality check.

MT Road Test

sting ray at a glance...

Easily the most powerful production car made... designed and built as an all-out sports car, yet fairly suitable for everyday use... stiff riding suspension makes around-town jaunts and long trips uncomfortable on the torso, but free of handling problems... styling exhibits a "going" look even when standing still... practiced drivers can feel safe at all speeds and in all conditions, due to 4-wheel disc brakes, quick steering, and good stability.

how the car performed...

ACCELERATION (2 aboard)
- 0-30 mph....2.5 secs.   0-60 mph....5.5 secs.
- 0-45 mph....3.8 secs.   0-75 mph....7.5 secs.

TIME AND DISTANCE TO ATTAIN PASSING SPEEDS:
- 40-60 mph .... 2.1 secs., 151 ft.
- 50-70 mph .... 2.0 secs., 172 ft.

STANDING START QUARTER-MILE:
13.8 secs., 104 mph.

BEST SPEEDS IN GEARS @ SHIFT POINTS:
- 1st .... 65 mph @ 6500 rpm
- 2nd .... 87 mph @ 6500 rpm
- 3rd .... 112 mph @ 6500 rpm
- Top Speed .... 143 mph @ 6500 rpm

MPH PER 1000 RPM: 22.0

SPEEDOMETER ERROR:
- Calibrated Speedometer .30 45 50 60 70 80
- Car's Speedometer 28 44 49 58 67 77

STOPPING DISTANCES:
From 30 mph, 31 ft.; from 60 mph, 135 ft.

specifications...

ENGINE: Ohv V-8
- Bore and stroke (ins.): 4.251 x 3.76
- Displacement (cu. ins.): 427
- Max. torque (lbs.-ft.): 460 @ 4000 rpm
- Horsepower: 435 @ 5800 rpm
- Compression ratio: 11.0:1
- Carburetion: 3 2-bbl. Holleys

TRANSMISSION: Optional 4-spd. manual, floor-mounted lever, all forward gears fully synchronous. Ratios of 2.20 in 1st, 1.64 in 2nd, 1.27 in 3rd, and 1.00:1 in 4th gear.

FINAL DRIVE RATIO: 3.55:1

SUSPENSION: Independent front, single lateral arm-type with coil spring, tube shock and spherically jointed steering knuckle at each wheel. Fully independent rear with fixed differential, transverse multi-leaf spring, lateral struts and universally jointed axle shafts. Tube shocks at each rear wheel.

STEERING: Optional linkage-type power assist, with semi-reversible recirculating ball nut. 17.6:1 overall gear ratio. Turning diameter: 39.9 ft., curb-to-curb. Number of turns lock-to-lock: 2.92

WHEELS: Short spoke disc steel, 15-in. dia. x 6-in. wide.

TIRES: 7.75 x 15 rayon, standard.

BRAKES: 4-wheel hydraulic, caliper disc actuated by dual system. Power optional. Dia. of disc front and rear: 11.75 ins. Effective lining area: 78.1 sq. ins.

FUEL CAPACITY: 20 gals.

MILEAGE RANGE: 9.0 – 12.0 mpg

BODY AND FRAME: Separate construction. All-welded, full-length, ladder-constructed frame with 5 cross-members. Fiberglass body.

DIMENSIONS: Wheelbase: 98.0 ins. Track: front 57.6 ins., rear 58.3 ins. Overall length 175.1 ins., width 69.6 ins., height 49.8 ins.

USABLE TRUNK CAPACITY: 8.1 cu. ft.

CURB WEIGHT: 3340 lbs.

prices and accessories...

MANUFACTURER'S SUGGESTED RETAIL: (includes federal excise tax but excludes state and local taxes, license, options, accessories and transportation) Soft top Corvette $4327.50 (with 327, 300-hp V8).

OPTIONS & ACCESSORIES:
- 435-hp, 427-cu.-in. V-8 .... $437.10
- 4-speed transmission 184.35
- Positraction axle (limited-slip) 42.15
- Red Stripe, special nylon tires 46.65
- Special cast aluminum wheels 263.30
- Power steering 94.80
- Power brakes 42.15
- Heavy-duty brakes 342.30
- Side-mounted, off-the-road exhausts . 131.65
- Full-transistor ignition 73.75
- Special-purpose suspension 36.90
- 36-gal. fuel tank (coupe only) 198.05
- Shoulder belts 26.35
- Power windows 57.95
- Removable hardtop (with soft top) .. 231.75
- Air conditioning 412.90
- AM-FM radio 172.75

---

gt 500 at a glance...

Eye catching styling is product of good judgment in restyling already good looking Mustang fastback... well suited for street duties, but refinements needed to make it an all-out champ on the track... one of the most comfortable cars for touring we've driven, but in-and-out from driver's seat is a tight fit... fiberglass additions would benefit from tighter quality control inspection... shoulder harness impressed us to the point of dissatisfaction with all others.

how the car performed...

ACCELERATION (2 aboard)
- 0-30 mph....2.9 secs.   0-60 mph....6.2 secs.
- 0-45 mph....4.4 secs.   0-75 mph....9.5 secs.

TIME AND DISTANCE TO ATTAIN PASSING SPEEDS:
- 40-60 mph .... 2.6 secs., 190 ft.
- 50-70 mph .... 3.0 secs., 264 ft.

STANDING START QUARTER-MILE:
14.52 secs., 101.35 mph

BEST SPEEDS IN GEARS @ SHIFT POINTS:
- 1st .... 51 mph @ 5500 rpm
- 2nd .... 68 mph @ 5500 rpm
- 3rd .... 93 mph @ 5500 rpm
- Top Speed .... 120 mph @ 5500 rpm

MPH PER 1000 RPM: 21.1

SPEEDOMETER ERROR:
- Calibrated Speedometer .30 45 50 60 70 80
- Car's Speedometer 27 40 45 55 64 72

STOPPING DISTANCES:
From 30 mph, 40 ft.; from 60 mph, 144 ft.

specifications...

ENGINE: Ohv V-8
- Bore and stroke (ins.): 4.13 x 3.984
- Displacement (cu. ins.): 428
- Horsepower: 355 @ 5400 rpm
- Max. torque (lbs.-ft.): 420 @ 3200 rpm
- Compression ratio: 10.5:1
- Carburetion: 2 4-bbl. Holleys

TRANSMISSION: Fully synchronized (forward gears) 4-spd. manual. Floor-mounted lever. Gear ratios: 1st 2.32, 2nd 1.69, 3rd 1.29, 4th 1.00:1.

FINAL DRIVE RATIO: 3.50:1

SUSPENSION: Independent front with coil spring and ball joints, modified for flatter cornering. .94 in.-dia. front stabilizer bar. Straddle-mounted H-D rear axle, single unit, suspended with longitudinal 4-leaf springs with special rebound dampers to control rear spring wind-up. Preset, adjustable shock absorbers at each wheel.

STEERING: Recirculating ball and nut, linkage-type power assist standard. 16-to-1 overall gear ratio. Turning diameter, 37.16 ft. curb-to-curb. Turns lock-to-lock, 3.74.

WHEELS: Shelby steel wheel, 15-in. dia. x 6.5-in. wide.

TIRES: "Speedway 350" Goodyear low-profile 4-ply nylon E70-15 140-mph rated tires, standard.

BRAKES: Dual-system hydraulic. Front disc/rear drum with power assist and high-speed linings on disc caliper, standard. Dia. of front disc: 11.3 ins. Dia. of rear drum: 10 ins. Effective lining area: 191.0 sq. ins.

FUEL CAPACITY: 16 gals.

MILEAGE RANGE: 7.6 – 14.6 mpg

BODY AND FRAME: Platform-type unitized construction with reinforced floor side-members and export front end reinforcement.

DIMENSIONS: Wheelbase: 108.0 ins. Track: front 58.0 ins., rear 58.0 ins. Overall length 186.6 ins., width 70.9 ins., height 51.6 ins.

USABLE TRUNK CAPACITY: 5.1 cu. ft.

CURB WEIGHT: 3360 lbs.

prices and accessories...

MANUFACTURER'S SUGGESTED RETAIL: (includes federal excise tax, but excludes state and local taxes, license, options, accessories and transportation) 2-door fastback Shelby GT 500 Mustang — $4195.00 (f.o.b. St. Louis, Mo.)

OPTIONS & ACCESSORIES:
- Mandatory options: power brakes, power steering, shoulder harnesses (2) $200.00
- Fold-down rear seat (mandatory) 64.77
- C-6 Cruise-O-Matic 50.00
- Air conditioning 356.09
- AM radio 57.51
- AM-FM radio 135.00
- Styled steel wheels 185.00
- Rally stripes (over the top) 24.95
- Limited-slip differential 181.00

HOTTEST 'VETTE YE

by Eric Dahlquist *With triple carbs, solid lifters, special cam and 4*

We're going out to the Riverside 500, see — part two since it rained the week before. And we know this neat little cutoff that knocks about five minutes off your ETA at the track. So we slip into the lane for the secret off-ramp and there's a line! Ordinarily this would be unusual by itself because hardly anyone uses the road, but in this instance it's almost an event — all the machines are Sting Rays! Well, OK, we see a checkpoint farther along and realize one of the local Corvette clubs is having a rally of some sort. The whole business kind of fades back to a corner of our mind reserved for Corvette rallies until we get near the track; a whole line of Sting Rays are cruising along, wending their way to the race. But most of these cars are something else, 427's with big Indy boots on mag wheels, those expensive Corvette mag wheels, and the exposed rocker panel exhaust setup that was introduced when they first slung the 396 in the Sting Ray and that is supposed to be used in off-road applications only. But most of all, there were adaptations of that wild two-tone hood paint treatment that came out with the '67 fake air-scoop hood. One of Rommel's Afrika Corps scout cars couldn't get any more looks that these jobs. Sting Rays are popular? We guess.

Only it wasn't supposed to happen that way. Back in 1963, when the new Corvette came out, the Sting Ray was one of the most radical designs in production — kind of the way you always thought the Batmobile ought to look, George Barris notwithstanding. Unfortunately, the sneaky old Jaguar people had brought out their swoopy XKE coupe several months before, which kind of took the edge off a styling scene that was just waiting around for something to happen. The base price differential on the two cars was about a grand, but this was not quite a true picture because by the time you got some of the options you wanted on the 'Vette that were standard on the XKE, the economy gap was shaved down to a point where it only meant a couple of extra payments.

Hands down, the Corvette was a more practical car. All the driveline pieces were sturdy and relatively inexpensive to replace if anything did go wrong; with even a 4-bbl Rochester on the 327, it would drop the British interloper with Matt Dillon-like regularity — and the best of all, that beautifully-shaped fiberglass wouldn't ever rust. And if you don't think the corrosion factor is a big problem in places like Detroit, Cleveland or Buffalo, just ask any local resident about those little brown decals that start appearing on rocker panels and rear quarters after a few years on the salt. As far as handling went, the XKE couldn't hold onto the tail of a well-prepared independent rear suspension Sting Ray, proving that Detroit could build extremely roadable cars when it had the opportunity.

The big handicap that most enthusiasts saw in those days was the basic drawback of all mass-produced things — there would just be too many of them. "In five years people will still turn to eye an XKE, but a Sting Ray will cause hardly a glance." For a time it looked as if this might be true, but with Zora Duntov back there in the office, we should have known better. The IRS, disc brakes on all four wheels, jazzy rocker panel exhausts, the 396's and 427's with virile bulging hoods, and this year triple carburetion and 435 horses — oh boy, what next?

The fact of the matter is that the culmination of all the Sting Ray "pizazz" in the 435 model has created a backlash — it has out-statused the XKE. Ask any kid you meet what the hottest thing going off the showroom floor is and you get one answer: a 435 'Vette. Almost overnight, nearly every Sting Ray you see is a tri-carb model. And these aren't stripped cheapies, either; they're in the Lincoln and Caddy price range. The guys at the insurance counters will all tell you that.

So, being observers of the latest social movements, we got programmed into a keen little 435-hp convertible with a fuzz-catching yellow and black paint job, Muncie 4-speed box (2.20:1 low) and 3.55 "Posi" rear end. Not so keen were a set

photography: Eric Rickman

LEFT — Everything's happening on the inside. Still true, this best 'Vette yet has sufficient gauges to suit any afficionado — aircraft or sports car. And how about interior room? Oh, yes, there's aplenty. Plush carpeted area behind the front buckets offers only .2 feet less luggage space than Camaro — with top up that is. With top down? Oh, well!
BELOW — Here's what the Sting Ray is all about. Even with non-adhesive skins, svelte roadster was superb.

sepower, here's a stocker that storms right off the showroom floor

HOTTEST 'VETTE

of puny-looking U.S. Royal "Laredo" tires (7.75 x 15) that somehow didn't match the rest of the car's masculine flavor. We just knew they were going to burn up as soon as we dropped the hammer (clutch) with anything but a closed throttle, and they didn't disappoint us. The Corvette's cockpit has the specific gravity of a P-51 Mustang, and that's the way it should be. No end of pleasure can be derived from watching all the gauges and needles doing their stuff every time you fire up, which is a good thing, aside from the fact of being knowledgeable about what's going on under the hood.

And there's a lot to be knowledgeable about. Some people felt that dropping in the 396 was stretching things a bit, but you can't even get that anymore — only 327's and 427's now, baby. The spec's on our engine were straight out of racing-land: 11.0:1 compression; .5197-inch lift solid-lifter cam; 2.195- and 1.725-inch-diameter intake and exhaust valves; 3 Holley carbs. About the last thing you'd expect to see in such a power factory is a horsepower-robbing smog pump, but there was one sitting

ABOVE — Over the years since its inception, Corvette styling has changed just enough to make it seem as fresh as ever. With discs on all fours, 'Ray is still the only mass-produced machine to have them, and still the best stopping. BELOW — You hear this rump-rump and you know the king is here.

VEHICLE	
Corvette Sting Ray	
PRICE	
As tested	$5,350.00
ENGINE	
Cylinders	8
Bore and stroke	4.251 x 3.76
Displacement	427 cu. in.
Compression ratio	11.0 to 1
Maximum horsepower	435 @ 5800 rpm
Maximum torque	460 @ 4000 rpm
Valves: Intake	2.195 in.
Exhaust	1.725 in.
Camshaft:	
Lift	.5197 intake, .5197 exhaust
Duration	316° intake, 302° exhaust
Carburetion	3 Holley 2-bbls.
Exhaust system	Dual 2.50-in. exhaust pipe, 2.00-in. tail pipe
TRANSMISSION	
Type	Muncie 4-speed
Ratios: 1st	2.20
2nd	1.64
3rd	1.27
4th	1.00
DIFFERENTIAL	
Type	Semi-floating, overhung pinion
Ring gear diameter	8.375 in.
Ratio	3.55 to 1
BRAKES	
Type	Disc
Dimensions: Front	11.75 in.
Rear	11.75 in.
Swept area	461.2 in.
SUSPENSION	
Front	Independent coil
Rear	Independent, multi-leaf
Stabilizer	.875-in. diameter
Tires	7.75 x 15
Rims	6-in.-wide
Steering gear:	
Type	Power
Ratio	17.6 to 1
Turning circle	39.9 ft.
Turns of steering wheel, lock to lock	2.92
PERFORMANCE	
0-30	2.5 sec.
0-60	5.5 sec.
Standing quarter-mile	108 mph in 13.80 sec.
DIMENSIONS	
Wheelbase	98.0 in.
Front track	57.6 in.
Rear track	58.3 in.
Overall height	49.8 in.
Overall width	69.6 in.
Overall length	175.1 in.
Curb weight	3340 lbs.
Crankcase capacity	5 qt.
Cooling system	22 qt.
Fuel tank	20 gal.

there big as life. On account of this, carburetor calibrations were a bit on the lean side and cold starting often consumed about 5 minutes of start-stall gyrations before the big 4.251 x 3.76-inch mill was merrily ticking over.

You don't really want to whack this machine away from the curb with vigor because you're just liable to find yourself in a big brody — the rubber is not what one would call bitey. And, as fast as the car will go with no effort at all, there are a few things to make right before quarter-mile assaults. To begin with, the front and rear carburetors are vacuum-operated (i.e., they come in when the air rushing through the ven-

HOTTEST 'VETTE

turis is sufficient to overcome a diaphragm/spring arrangement). Additionally, to smooth general carburetor action, the operation is made less responsive by a check-ball so that the butterflies will not slam shut during a shift on a hill, say. The drag racer (after mounting some decent tires) will want to quicken this operation by removing the check-balls altogether and reducing spring tension in the diaphragm by cutting off some of the coils so the carbs come in quicker. The best method is to cut about a quarter-coil off at a time until carburetor action is what you want.

The second point is the diaphragm pressure plate. The first time we tried a 1-2 power shift, the pedal went down and stayed there. The problem is that centrifugal force keeps the pressure plate disengaged. When Chevy was in racing back in '62, they had the same problem, but instead of tearing out the diaphragm like most of the fellows did, the solution is in adjusting the pedal down to where it will just disengage, or by putting a block of wood under the clutch pedal so you can't push it over center. With these things done, plus removing all the accessory drive belts, airing the tires to 36 pounds and ballasting with a full tank of gas, the Sting Ray ran a good 13.80 – 108, even then severely handicapped by lack of traction. We say again, the one thing the car needs is better skins — period.

Another consideration might be a better ride at regular speeds. When they dropped in the heavy 396-427 engine, one of the penalties imposed by the swap was higher spring rates. As such, the 427 model is strictly a smooth-road machine at the posted speed limits. Granted, once you get wailing, the suspension evens out and sticks to the ground doing it, but there are few places left to run a hundred-twenty for sustained periods.

As you might expect, steering under all conditions was quick and light, and though power-assisted, provided an excellent feeling of changing road surfaces. With a lot of the weight hung on the front end and all that awful awful power, the car tends to dart some when a new driver slips behind the tiller, but a little road time will cure this in a hurry. On very tight turns, we found the 427 tends to plow slightly, a situation that can be remedied immediately by adding throttle to bring the rear end around and permanently, at least in part, by the substitution of, again, some higher level tires. We know a fellow journalist who has one of these bombs outfitted with radial plys, and he claims this is the answer as far as he's concerned. No matter what you do with the car, it always comes back to the same thing; the ride is just too severe for any kind of protracted driving, which is a real shame because the Sting Ray's other attributes — steering, balance, adequate leg room, good seat-to-steering-wheel relation, disc brakes that are far superior to anything else we've tried and, of course, the spine-snapping response — are just the right ingredients for a Grand Touring car in anybody's language.

From a visual standpoint, that yellow lacquer is the car's most arresting aspect, and it does a pretty good job at being smooth and free of orange peel. Early Sting Rays endured a session of poor-fitting fiberglass components, a situation we thought had been put behind years ago but appears to crop up in places, like the way our doors failed to match the rear quarters by almost ½-inch. With the top of the convertible in place, luggage capacity is 8.1 cubic feet, only fractionally less than the Camaro and Mustang, and the Corvette doesn't even pretend to be utilitarian. The thing that sells the 435 Sting Ray is that, like many Corvettes of the past, for this time and place, it is the hot setup. A car that an owner can have fun with, get startling good gas mileage considering the tri-carbs (11-13 mpg) and not have to spend hours tinkering on. That, in a nutshell, is the problem with Chrysler's street hemi — when it's razor sharp it's great, but when it's out of tune it's terrible. The Corvette, like many of the Corporation's hot machines, will go on and on, shutting down would-be challengers with minimum maintenance. GM may not be in racing but its divisions build the best darned line of production competition cars in the world. The 435 Sting Ray is kind of king of these kings.

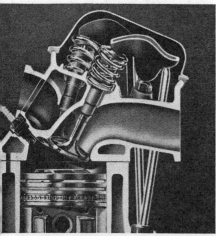

ABOVE — Oh, the fuel mixture goes in here, and it comes out here. And while it tumbles and turns in the semi-hemi combustion chamber, the atmosphere gets very close, very close. 11.0:1 maybe.
BELOW — Fastback coupe is most popular model, has more space inside and with the normal suspension is one of the nicest handling, fastest cars around.

THE STING RAY EMERGES

LAUNCHING A NEW ERA IN THE PLASTIC SPEEDSTER'S EVOLUTION.

BY SPENCE MURRAY

Then as now I was a member of the motoring press and as such, had been exposed early to illustrations depicting the coming new Corvette Sting Ray models, although I'd missed the press preview of them at General Motors' Milford Proving Grounds in June of 1962. While I had been a Vette admirer since the unveiling of the original Motorama Show roadster at New York in 1953, my role as a family man left no room in the budget (or garage) for a 2-passenger personal car. But one look at GM's photos of the yet-to-be-announced Sting Rays found me overly entranced by what would later come to be called the famed split-window coupe. I simply had to have one.

Even before public announcement of the Sting Rays, I pleaded with my local Chevrolet dealer to accept my order for one and I went tip-toeing through the option list to specify how I wanted the car equipped. I chose carefully to keep the (for me) already high base price within reason, then went home hoping my early order would be processed before the expected rush of buyers became unmanageable.

Several days prior to public unveiling of the all-new Vette, my dealer telephoned; he had just received an unexpected delivery of four Sting Rays; three coupes and one convertible, and he was parceling them out on a first-come, first-serve basis against his backlog of orders which by now was sizeable. The convertible and two of the coupes had already been accepted but the fourth one was refused by its would-be buyer since it did not incorporate the specific options he wanted. Although the car was not equipped even close to my order either, I could take imme-

1

diate delivery although with the provision that I return the car to the dealership for each of several following weekends for show-room display.

Although Chevrolet had prepared 25 pilot, virtually hand-assembled Sting Rays for advertising and other media photography, for press corps test driving and, later, for special shows and displays, as early as June, production start-up didn't begin until October of '62 and even then, the flow was only the merest of trickles at first due to complex assembly logistics. Four cars were taken from the first few days' dribble and shipped to California for dealer showings and service familiarization. Plans changed, however, while the car-carrying transporter was en route from the factory and the cars somehow ultimately arrived at my dealership. My coupe had been the last one loaded at St. Louis, thus it was the first off-loaded and the first production Sting Ray to touch California soil.

THE REAL THING

Within the hour I saw my first real live Sting Ray and the car looked even better to me than it had in the sometimes deceptive photography. I quickly signed my acceptance of the Daytona Blue, 300 hp, 4-speed equipped split-window, but not without a lot of soul-searching on having to face what seemed at the time outlandishly high monthly payments. All told, with options, tax, licensing and dealer preparation, the coupe's price tag nudged the $5500 mark—nearly twice that of other Chevrolets I had bought in the past.

Of the three other cars on that first-to-California load, two fell in with prestigious company; one for actor/singer Sammy Davis, Jr., and another for Wally Parks, then as now president of the National Hot Rod Association.

A little over five years later, in mid-'68, I wrote a follow-up on that same Sting Ray for a magazine and shall excerpt from it here:

"It took me 1,932 days to drive my Corvette 100,000 miles, but in so doing some interesting statistics came to light that might be worth passing on:

"The '63 fastback runs the 300-hp, 327 engine with single quad, 4-speed box and limited slip rear-end with 4.11's. It's been used almost daily over the 5-plus years, except for periods of some inactivity or when laid up for repairs. During the 63 months, though, it has averaged 52 miles of use per day. When new it produced a low, low 7.5 mpg in traffic, but up to 17 mpg on the open road. In the end its overall mileage works out to 15.59 mpg—not bad considering that gas-hungry engine and low differential ratio. However, I had an outlay of $2,432.31 which purchased 6,410.3 gallons of premium gas.

"Repairs to the Vette were minimal during its first year—only $53.05—but the cost sheet shows a steady upward trend year by year (with almost identical miles of annual use) until the final tally came to $1,621.80—including three sets of tires, three paint jobs (by Joe Bailon, George Barris and Dave Puhl), plugs, points, etc., every 10,000 miles, oil changes, service, and so forth. All told, the car has cost $4,207.94 to operate which does not, of course, include its original price, annual licensing, insurance, etc. This sounds like a lot of money, and it is!—but computation shows a 4.2¢ per mile operating cost; within reason, all things considered."

That split-window remained in the family for two more years and, alas, when finally disposed of in late 1970 (traded even-up for a Sunbeam Tiger) it showed 140,000 miles on the odometer. During all this time the engine was never once "opened up" and I had experienced no greater service problems than two clutch replacements. Hindsight being what it is, I should have put that early coupe up on the proverbial blocks and let my eldest son do without the Sunbeam.

START-UP

Mindful that production start-up on the new Sting Rays might come acropper, Chevrolet was necessarily hesitant to announce a public

1. It has been GM's history that when they create a unique styling treatment, they heavily indulge themselves in promoting it. So it was when the car lines went to '55's wraparound windshields, and they followed suit with the famous Sting Ray split-window. This photo and others similar to it, were splashed in ads, posters, and sales brochures throughout 1963. This prototype car carries fuel injection as well as the new knock-off wheels.

2. There was early dissension in the ranks over the sales potential between the coupe and the roadster. It turned out, though, that while the coupe came on strong at first—likely due to the annually inclement winter weather with a closed car preferred, as well as to the uniqueness of the body—roadster sales later in the year caught up. Final tally: about 10,900 roadsters (half with hardtop) and some 10,600 split-windows.

THE STING RAY EMERGES

introduction date for the '63 convertible and coupe. Body assembly problems for the fiberglass sportster, especially in its fastback configuration, might be expected to delay full-scale production until well after the first of the year. Motor Trend Magazine, ever with its ear to Corvette's ground, reported with authority in its issue for October, 1962:

"Don't look for the new corvette fastback coupe next fall with the 1963 models. Prototypes have been running for several months, but word is that this radical body will be held back for a "1963½" introduction next spring. It should then make almost as big a splash as bringing out a complete new car. The new coupe body is said to be stunning, making the '63 convertible look classic by comparison. However, it will not be a four-seater, as has been rumored. The rear compartment is used for storage. And there is no trunk lid as such. Access to the storage area behind the seats is through the side door and over the seat backs. A trunk lid would have broken up the beautiful lines at the rear. This one will be worth waiting for."

It cannot now be determined if this notice was a "plant" from Chevrolet as a hedge against possible assembly-line problems, but it is certainly more than mere speculation on the part of the writer. Notice the reference to the fact that the Sting Rays would not have an opening trunk. Had this been merely a guess, then this stand would not have been taken for one of the later prototype cars built before final production "freeze" did in fact include an openable decklid—on the order of today's hatchback sedans and station wagons wherein the swing-up hatch includes the rear window!

Nevertheless, and against what must have seemed at the time insurmountable odds, convertible and coupe Sting Rays did begin rolling from St. Louis, and on a two-shift basis, as originally scheduled. In their very next issue, Motor Trend (for November) featured in-depth looks at both body styles stating the cars were in dealer showrooms. Photography for this edition had been taken of pilot cars used at the Milford press conference in June and they editorially lauded the handiwork of Bill Mitchell and Zora Arkus-Duntov and their respective engineering staffs.

Thus was the Sting Ray launched, and on a largely unsuspecting audience although glimpses of what had been going on at Chevrolet had long been hinted at through the motoring media. Early orders quickly outstripped production rates and by the first of 1963 would-be Corvette buyers were cautioned that a waiting period of at least 60 days existed.

EARLY ROAD TEST

Those first Sting Rays were in such demand that few were funneled out of the mainstream for roles as press familiarization and test cars. Add to this Chevy's undoubted reluctance to let the always critical, often hard-driving magazine and newspaper writers wring out early-run cars where assembly and service problems may arise sooner than on cars built later in the mod-

el run. A new Sting Ray coupe thus wasn't available for magazine road testing until well after the first of 1963. And, publication production time being what it is, Motor Trend's first full-fledged road test of that fuel-injected fastback did not appear until the magazine's May issue. This is what the editors had to say at that time:

"For the first time in its 10-year history, the Corvette is in such demand that the factory has had to put on a second shift and still can't begin to supply cars fast enough. The waiting period is at least 60 days, and dealers won't "deal" a bit on either coupes or roadsters. Both are going for the full sticker price, with absolutely no discount and very little (if any) over-allowance on trade-ins.

"This is a healthy situation for Chevrolet, and we're happy to see the 'Vette get the public acceptance we've always felt it deserved. Yet, after giving the new one a thorough shakedown, we can't help but let our thoughts stray back to last year's road test and one statement in particular we made.

"It had to do with how the factory has never really made any big profits on the Corvette, but that Chevy brass was more than satisfied as long as it carried its performance image and prestige over to the bread-and-butter lines. We also ventured an opinion that as long as the factory kept building the car on this basis it would be a great automobile, but if they ever put it on a straight dollar-profit basis, the Corvette would probably be ruined.

"Well, they haven't ruined it yet, but our test car (as well as several others we've checked out) showed definite signs that the factory might be getting more interested in dollars than in prestige. From the important styling and all-around performance angles, the new Sting Ray is an even greater car than its predecessors. But for a car that sells in the $4500-6000 range, it doesn't reflect the degree of quality control we feel it should.

"To begin with, there still seems to be some difficulty in manufacturing a really smooth fiberglass body. While this isn't too apparent in a light-colored car, it becomes all too noticeable in some of the darker ones. When the light hits these from almost any angle, there's a definite rippled effect. The doors on our test car had gaps around them that were wider than they should have been. The doors didn't line up too well, either. This was also true of the hood.

"In the past, we've always been impressed with interior trim in most GM products. The moldings usually look as if they've been very carefully designed for a precision fit. To our eye, the Sting Ray coupe's interior had an unfinished look. Not that the upholstery and carpet materials weren't top grade—they were—but the various door and window moldings aren't too well designed in the first place, and it doesn't take much laxity of quality control on the assembly line to make them look really bad. While these aren't earthshaking faults or defects, and have absolutely nothing to do with the operation of the car, they are of the sort that a discerning owner and driver will be constantly aware of.

"For the first time, the Corvette is available with power steering and power brakes. We didn't have either on the test car, but we've driven several set up this way. Combined with one of the smaller engines and Powerglide transmission, these power accessories make the Sting Ray docile enough for little old ladies or any other types interested in a nice, quiet, Sunday-go-to-meeting car.

"The basic power trains are carried over from last year and include four engines of 250, 300, 340, and 360 hp. Powerglide's available with either the 250- or 300-hp engines. Basic transmission with all engines is a three-speed manual, with four-speed available optionally. The 340- and 360-hp engines use the close-ratio (2.20-to-1 low gear) four-speed. Six rear-axle ratios are available and include the 3.36 (standard with three-speeds and Powerglide), 3.70 (standard with four-speed transmission), plus other gearsets of 3.08, 3.55, 4.11, and 4.56 to 1.

"The Motor Trend test car was equipped with the fuel-injection, 360-hp mill, four-speed, and 3.70-to-1 Positraction rear axle. Sintered-iron brake linings and heavy-duty suspension completed the option package.

"On a straight acceleration basis, there's very little difference between last year's car and the new one. Our quarter-mile times are within fractions of what they were last year. The only real difference is that the new one doesn't have quite the wheelspin (with stock tires) that the old rigid-axle car had. The 0-30, 0-45, and 0-60-mph steps averaged 2.9, 4.2, and 5.8 seconds, while our average time through the quarter-mile traps was 102 mph, with a 14.5-second ET. Top speed was an honest 130 mph,

1. The lithe look bestowed on the new Sting Ray Corvette was helped by the "nipping in" of the body at the rear of the doors. The roadster/hardtop and, especially, the coupe, exhibited styling advances unique for a production car in an era of standstill throughout the automotive industry.

2. The convertible found nearly equal favor among Corvette devotees, many probably opting for it due to the "strange" appearance of the coupe's split rear window. Of the 10,000-plus converts, choice between soft-top and hardtop was about evenly divided.

THE STING RAY EMERGES

with the tachometer reading 6000 rpm. A course longer than the Riverside Raceway backstretch would've produced something very close to the Sting Ray's theoretical top speed of 140-142 mph (with 3.70 gears), because the engine was still winding when we had to back off. The 360-hp engine is set up like any well-designed racing engine, and it's very strong throughout the entire rpm range.

"This is one of the few high-output engines that can deliver decent gas mileage without being babied. Out on the highway, we averaged slightly better than 18 mpg for one trip where we didn't go above the legal limits. On another trip where the speedometer stayed above 75 and 80 mph a good deal of the time, we saw 16.3 mpg. Whipping around town produced a 13.6-mpg average. For over 700 miles of all types of driving, the Sting Ray averaged 14.1 mpg.

"A lot of this is due to the low weight the engine has to pull around, but the excellent Rochester fuel-injection unit also has something to do with it. Like any good injection unit, it can constantly correct itself to suit different humidity, temperature, and altitude conditions. All in all, it's a completely efficient system. It doesn't seem to be temperamental, either. Care does have to be taken to keep the fuel filters clean and operating because they're quite susceptible to dirt.

"We'd recommend the metallic brake option to any buyer, regardless of the engine he's getting in his new Corvette. It's very reasonably priced ($37.70), and is unmatched for efficiency. The brakes in the test car were used very hard at the end of several high-speed runs and showed very little tendency to fade. It's true that they require more pedal pressure to operate and are a trifle noisy on cold mornings, but once they get warmed up they're excellent. If a woman is going to be driving the car a lot, power assist can be added to keep her from developing bulging muscles in the brake leg. Self-adjusting brakes are now standard equipment, and unlike systems in most other cars, they'll adjust as the car moves forward. The brakes in our test car pulled the Sting Ray down to quick straight-line stops time and again without any sudden locking of the wheels and without apparent fade. Several stretches of mountain roads showed that they could stand up to prolonged hard use without failure.

"For all-out competition there's a special performance brake option that includes a little more effective brake lining area, bigger finned drums, power assist, and a divided output master cylinder (separate system for front and rear), and provisions for cooling. Combined with the optional cast-alloy wheels, this setup gives the competition Corvette braking power on par with many of its disc-braked competitors.

"The new suspension system is far in advance, both in ride and handling, of anything now being built in the United States. It's completely comfortable without being mushy, and it takes a large chuck hole to induce any degree of harshness into the ride. Sudden dips, when taken at speed, don't produce any unpleasant oscillations, and the front and rear suspension is very hard to bottom. There's very little pitch noticeable in the ride, even though the Vette is built on a fairly short (98-inch) wheelbase. At high cruising speeds—and even at maximum speeds—nothing but an all-out competition car will equal it in stability. We drove it under some pretty windy conditions and didn't notice any adverse effects from crosswind loading.

"We thought the old model cornered darn well, but there's no comparing it to this new one. It does take a little different technique, but once the driver gets onto it, it's beautiful. Since the 49/51 per cent front-to-rear weight distribution, plus the independent rear suspension, gives the Sting Ray an inherent amount of oversteer, the driver will find that on fast corners the car will be doing most of the work through the corner instead of him powering it through.

"At most speeds the coupe and the roadster are quite noisy. In addition to high engine and wind noise levels, the coupe picks up and amplifies quite a bit of road noise through the differential, which is rigidly mounted (although in rubber) to the frame. The extremely stiff, ladder frame and well-designed body, with its built-in steel bracing (on the coupe), keep body shake to an absolute minimum.

"The Sting Ray is roomier than the older models, and quite a bit of luggage can be carried in the space provided. Some people will be unhappy without a deck lid, but it's really not too inconvenient to get to the storage space through the passenger compartment. The steering wheel is now adjustable, although not readily so from the driver's seat. It has to be done in the engine compartment with a simple wrench, but gives up to three inches of fore and aft adjustment.

"The bucket seats offer more of a contour fit to the back and are very comfortable once you get used to the low seating position. A full array of instruments is functionally arranged on the dash within easy view of the driver's eyes. But their design is such that at night, with the instrument lights on, they're hard to read. The brushed aluminum backing of each gauge tends to glare. The rear window on the coupe is designed more for looks than practicality, and any decent view to the rear will have to be through an exterior side-view mirror."

1. Layout of the independent rear suspension was an engineering challenge, ably met by Zora Arkus-Duntov, but equally as challenging was convincing the corporate powers that the cost of the rear-end assembly could be partially offset by using proprietary Chevrolet front suspension pieces. Critics found the multi-leaf spring rather archaic, but later determined road adhesion to be greatly advanced over the older, single-axle arrangement.

2. All prior Corvettes had the engine dead-centered in the frame, but the Sting Ray's powerplant was offset to the right an inch. This allowed true alignment between the crank center and the differential pinion shaft, thus permitting a narrower body tunnel and more comfortable foot room.

CHASSIS LAYOUT

Motor Trend was, obviously, largely impressed with the new Vette and while they underwent no little criticism for their somewhat unfair observations of body panel fits and molding details (their car was, of course, an early production model where assembly glitches are common), their remarks were considered highly laudatory. The magazine's evaluation of the new suspension system and general chassis layout was valid, for they delved editorially into the intricacies of chassis mechanics and the why's behind it, by no less than an authority than noted technical writer, Roger Huntington:

"For the last five years, we've been bombarded with rumors of an "all-new" Corvette that was supposed to be just around the corner. It was going to feature just about everything that was new and exciting in modern sports car design. We waited anxiously.

"There was no question about the demand for a new Corvette. In many ways the original basic chassis and body, in production with relatively minor changes since late 1953, were obsolete. It was just a question of how soon Chevrolet could afford a major tooling change—in view of the low production volume of 10,000 to 12,000 units a year. Only the naive hoped for an all-new Corvette every two or three years.

"Well, we finally have our all-new Corvette. The 1963 line features a brand new chassis from the ground up—new frame, completely new body (in both convertible and coupe)—and only the engine and transmission options remain from previous models. This is the one we've been waiting for. And it's all the rumors promised—and more. This is a modern sports car. In most ways it's as advanced as the latest dual-purpose sports/luxury cars from Europe—and this includes the new Jaguar XK-E, Ferrari GT, Mercedes 300-SL, and all the rest. The new Corvette doesn't have to take a back seat to any of them, in looks, performance, handling, or ride.

WHAT WERE THEY AFTER?

"Chevrolet engineers assigned to the Corvette project, headed by the well known Zora Arkus-Duntov, set down several basic design goals for the new car before they ever drew a line. One important aim was an improved ride. This may not seem vital in a sports car, but you've got to remember that the Corvette is designed to appeal to a larger market than just the purists and the racing enthusiasts. Many Corvette owners use the car as they would a Thunderbird—as a two-seater personal car. These people want a plush ride. The early Corvette wasn't bad in this department, but this new one is unbelievable for a car of its weight and wheelbase.

"A second basic aim was improved handling—roadability and cornering, both on the highway and on the race track. The early Corvette did a wonderful job in competition against all-out sports/racing cars. But its handling couldn't be described as "modern" or "advanced" in any sense of the words. There was too much understeer, or ploughing of the front end. You could wrestle it around a corner, but it wasn't an easy car to drive fast. One of Duntov's basic aims on this new chassis was to develop virtually neutral steer characteristics right up to the point of breakaway. In other words, there was to be no pronounced tendency for the front or rear end to wash out before the opposite end. He also wanted better wheel adhesion, or sticking, on rough corners and straight surfaces. The early Corvette left much to be desired in this area.

"A final basic aim (though not related directly to the chassis) was to design a body that looked good and yet had less wind resistance and better aerodynamic stability. This wasn't as easy as it sounds. It's traditionally been tough to com-

THE STING RAY EMERGES

bine good lines and aerodynamic efficiency in the same body. One area generally has to be compromised—and it's usually the efficiency. But Duntov also knew that many buyers would be racing their Corvettes, and unnecessary air drag and poor aerodynamic stability would hurt lap times at the speeds of 130 to 150 mph reached on some of these courses. So they came up with the beautiful new fastback coupe body. These goals have been achieved very nicely.

HOW THEY DESIGNED IT

"The general ride and handling characteristics of a given car are determined by many, many individual factors working together—and it's very risky to try to isolate any single factor as being the secret of a good-behaving car. But in the case of the new Corvette we must certainly mark down reduction in unsprung weight as a key factor in getting the desired effect.

"Unsprung weight is the mass that moves up and down with the wheels as they pass over bumps. This would include the wheel, tire, brake, axle, part of the spring weight, and so forth. The vertical motion of this unsprung mass is resisted by the sprung mass, or the weight that's actually supported by the springs (body, frame, etc.). Obviously the greater the sprung mass in relation to the unsprung mass, the less wheelhop we'll get, and thus the better the tire adhesion on rough surfaces. Also, the greater this sprung/unsprung weight ratio, the better the ride will be—simply because the sprung mass has so much more inertia than the unsprung that the vertical movement of the wheels can hardly budge the sprung mass, and it just glides over the bumps while the wheels go bobbing up and down. (If the sprung and unsprung weights are equal, the ride is terrible. Ever notice a small two-wheel car trailer with no load in it?)

"Unsprung weight is no problem on the front end of a modern car with independent front suspension. The wheels are carried on light arms, with a light coil spring usually between the frame and lower arm. And, of course, the heavy engine is on the sprung-weight side. But it's a different story at the rear end. Here we have a relatively light body weight, but with the heavy, "solid" rear axle bobbing up and down with the wheels. This combination is murder on ride and rear-wheel adhesion on rough surfaces. This was worse in the case of the early Corvette because, as car weight is reduced, the sprung weight has a tendency to decrease faster than the unsprung. That is, that early chassis used standard Chevrolet wheels, axle, etc.—and yet the Corvette's rear section weighed about 300 pounds less than the big Chevy. This made the sprung/unsprung weight ratio that much worse.

"The obvious answer was independent rear suspension—with the heavy differential mounted right to the frame (as sprung weight), and driving the wheels through U-jointed half shafts. This system has more unsprung weight than in front, because of the drive components and the extra beef required in the control links to resist engine torque. But it's a fat 33 per cent less than with the old solid axle as unsprung weight. Chevy engineers quote a rear-end unsprung weight of 301 pounds for the '62 chassis with solid rear axle. It's only 200 pounds for the new '63. The effect of this reduction on ride and wheel adhesion must be felt to be appreciated.

"Another important factor in ride is weight distribution, and the way the major masses are grouped in relation to the center of gravity. A

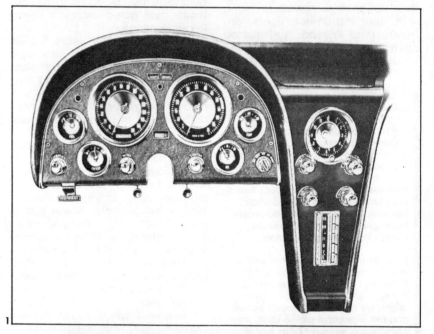

front/rear weight distribution of 50/50 is about ideal for ride. The '62 Corvette had slightly more than half its total weight (50½ per cent) on the front wheels with two people in it. The new car has a 49/51 distribution, with about 80 pounds more on the rear wheels. This is still near ideal for ride, and helps the traction noticeably.

"The main reasons for the greater rear-end weight is that the wheelbase has been shortened from 102 to 98 inches, which shoves everything back a little in relation to the rear wheels. Also, the single transverse leaf spring is situated directly behind the differential, which puts the effect of this mass back a few inches. This movement of major masses backward on the wheelbase has another beneficial effect: It increases the "flywheel effect" of these major masses about the center of gravity—which slows down the pitch rate of the car, and thus improves the ride. This, coupled with the reduction in unsprung weight, has given the new Corvette what seems like a Cadillac ride compared with the early design.

"The suspension layout on the new Corvette is quite interesting. Up front, things are very conventional—lateral wishbones of unequal length (with upper arm tilted to give anti-brake-dive geometry), coil springs, and tubular shocks inside the coils. In fact, many of these front suspension components are interchangeable with Chevy passenger cars. An anti-roll torsion bar is used to give added roll stiffness in front. A new recirculating-ball steering gear is used, of conventional geometry, with optional overall ratios of 19.6 (standard) and 17 to 1 for racing.

"The differential section is carried on two frame crossmembers, front and rear, and is mounted through thick rubber biscuits to isolate vibration and road rumble. This is a very clever solution to this problem. Vibration has always been a headache with frame-mounted differetials. Chevy engineers have attacked it by carrying the rear of the differential on a separate bolt-on frame crossmember, then bolting this to the main frame through these thick rubber pads. This crossmember wouldn't contribute much to frame stiffness, but it ought to be very effective in isolating vibrations. The front differential mounting bracket is also attached to a main crossmember ahead of the wheels through rubber.

"Springing is via a rather hefty nineleaf transverse leaf spring bolted to the back of the differential case, and linked to the outer wheel hub sections through rubber-cushioned pillions. (The layout is just like the cross-leaf front suspension on the old '36-37 front-drive Cord.) This flexible mounting of the spring to the hubs means that the spring carries absolutely no driving or cornering loads—unlike longitudinal leaf springs on a Hotchkiss rear end. All the spring has to do is to provide the vertical suspension effect. Suspension links take care of wheel motion.

"Some people question Chevy's choice of the "obsolete" multi-leaf spring on the new Corvette. Frankly, this was an economy move. The leaves are less expensive to make and assemble than the more complex coil-spring suspensions, with their extra brackets and fittings. Chevy engineers probably would've used coils or torsion bars on the rear if cost had been no object (in fact several early prototypes had coil rear ends); but the leaf spring was far more practical on the production line. Why didn't they use the more advanced single-leaf spring, like on the Chevy II? It would've been considerably lighter—and seems as if it would be cheaper. Chevy engineers won't say why they didn't.

"The rear suspension linkage geometry is unusual. In operation it's similar to an unequal-length wishbone front end. The U-jointed axle shafts themselves act as the upper control arms. This is possible because these shafts have no slip joints. In other words, they hold a constant length, and their effective pivot centers are at the centers of the U-joints at each end. Then the lower control arms are simple rods that pivot at the differential case on the inner ends and at the hub sections at the outer ends. These four "links" completely control the lateral motion of the wheels. Fore and aft motion, plus driving and braking torques, are resisted by box-section trailing arms fabricated from sheet steel, and pivot ahead of the wheels on the frame. Tubular shock absorbers are attached to the frame at the top and to the bottom of the hub sections.

"Rear suspension geometry is predetermined by adjusting the lengths, angles, and pivot points of these six control links. This is one of the keys to the good handling of this car. For example, the rear roll center (the point about which the sprung mass tilts in a turn) has been raised to 7.5 inches above ground level. This combines with a front roll center height of 3.25 inches (much higher than the conventional American passenger car) to give a relatively high roll axis. This, in turn, combines with the lower overall center of gravity of the car (3.3 inches lower than '62) to give a much shorter lever arm when centrifugal force tries to tilt the body about the roll axis in a turn. (This lever arm is the distance between the roll axis and the center of gravity.) This means a lower roll angle—which accounts for less weight transfer to the outside tires and less camber change on the wheels on hard corners. This new Corvette should show very little roll in the turns.

"A common fault of many link-type independent rear suspensions is that they have such a relationship between roll center height (usually very low), link positioning, and static wheel camber that the wheel assumes a very steep positive camber as the body rolls in a turn. That is, the top of the wheel tilts outward. This has the effect of increasing the tire slip angle, and increasing the oversteer tendency. Chevy engineers have completely solved this problem. They put 1.5 degrees of negative camber on their rear wheels at normal load. Then this changes to only 1.5 degrees of positive camber when the body is under full roll angle in a turn. Thus the rear wheels are virtually vertical through the whole suspension travel—which is ideal from the standpoint of getting neutral steer characteristics. In fact, it wasn't necessary to design any appreciable roll steer geometry into the rear suspension because this thing was so neutral right up to breakaway. (Roll steer is when the suspension linkage causes the rear wheels to steer slightly one way or the other when the body tilts in a turn or under a cross-wind force.)

FRAME AND BODY PROBLEMS

"No suspension system can work the way it's supposed to if you

---

1. It has been said, perhaps with some insight, that the dash layout of the Sting Ray was done separately using only the general body parameters. It is true that there is little unity between the double-humped dashboard and the car's exterior styling, but instrument placement was an improvement over the earlier Corvette.

2. Few early Sting Rays had the fit and finish of this one. It was one of four in the initial California delivery from St. Louis and intended for dealer familiarization and display (license tag reads 1962). In a 1963 issue, Motor Trend magazine remarked on the generally poor detailing on its test car.

THE STING RAY EMERGES

don't have a good, rigid frame to tie the four corners together. This "frame" can be either a good unit body or a basic sub-frame with body attached. It's well established that the welded unit body (actually a large box section) has more overall torsion and beam stiffness than most frame/body combinations. But, of course, the high tooling costs of such a body would rule it out for the low-volume Corvette.

"Also the fiberglass body on the Corvette (another necessity because of low volume) contributes somewhat less to overall frame/body stiffness than an equivalent steel body. So Chevy engineers had to really get down to business on the new frame for the '63. It had to have absolute maximum beam and torsion stiffness in relation to weight—and yet had to eliminate the old X-crossmember to give more convenient floor wells for the new, lower Corvette body.

"You can see the result. The new frame is a basic box configuration, with a very high kickup over the rear suspension and widespread side rails running under the outside edges of the body. There are five major crossmembers running straight across. Side rails are boxed for added stiffness. The new frame weighs almost exactly the same as the '62—about 260 pounds—but it has slightly more beam stiffness and 50 per cent more torsional stiffness. This increased torsional stiffness (resistance to twisting) should be noticeable to the racing boys on high-speed corners. This is bread-and-butter the same (or less) weight, more function from when you can get engineering at its best.

"Unfortunately, it's hard to reduce weight on a car that has a massive sub-frame like the Corvette. This is a big advantage of unit and "monocoque" construction. The new Corvette coupe weighs about 100 pounds more than last year's convertible—which would be a hair over 3400 pounds with full gas tank and two people. Maybe this is one of the prices for progress.

"The radical new fastback coupe body design was actually developed in the Cal Tech wind tunnel, using ⅜ scale models. GM stylist Peter Kyropoulos and his staff tried to get the best compromise between looks, wind resistance, and aerodynamic stability. The Chevrolet people aren't making any great claims for the result, but they do say that the total wind drag is well below last year's Corvette with hard top in place. And the drag coefficient itself is even less than that of the famous Sting Ray open competition car. (However, the frontal area would be more, so total air drag might be about the same.) Beyond this they won't release specific figures. We'll just have to wait to get some idea of the relationship between maximum speed and available horsepower on the fast race courses and at Bonneville.

"And that leaves only the important subject of brakes. They're better on the new chassis. They retain the same 11-inch drum diameter, but the front and rear lining widths are increased by ¾ and ¼ inch respectively—which increases the total effective lining area from 157 to 185 square inches. Segmented sintered iron linings are still optional for racing, but the new optional finned competition brake drums are now cast in aluminum, with cast-iron liners. This combination of metallic linings and aluminum drums should completely eliminate any possibility of fade under any conditions. Optional aluminum knock-off wheels will further aid cooling. It's certainly obvious that Chevrolet intends to continue to go racing with the Corvette."

The impact on the car-buying public of the Sting Ray's innovations was great, even in that time of generally high automotive interest and awareness. Of course, advance-concept styling, mechanical intricacies like independent rear suspension and disappearing headlights, and even such offerings as periscope-style rear vision mirrors, windshield wipers that answered automatically to rain, and so on, were being touted through the-then wave of Detroit dream cars of the early Sixties. People were just downright blase about automotive innovating, even though few offerings such as noted above had actually reached production status; that is, until the Sting Ray debuted. Here for the first time in post-war years was a for-real American production car with i.r.s., and disappearing headlights hadn't been seen since the long-ago demise of the 1942 DeSoto. Bill Mitchell's knife-edged styling sparked controversy—largely favorable—since U.S. offerings at the time were generally rounded and quite pointless, fins by now having gone the way of the dinosaurs. If the convertible, in either soft- or hardtop guise, was way-out, then the fastback coupe was light years ahead of anything else either at home or abroad.

THE GREAT SPLIT

One of Mitchell's styling touches, though, sparked a lot of comment, and not all of it was good. It was the coupe's split rear window; actually, a fully divided rear window, not merely "split" with a thin chrome strip as on cars of the Forties. GM's Chief of Styling had had a long-standing desire to use such a divider as a logical extension of a central, fore/aft character line. His chance came with the Sting Ray since the low hood necessitated a narrow central blister to clear the plenum of the optional fuel injection system. To him it seemed somehow right to extend the fillip back along the roof, down between the window halves, and on out to the abrupt duck's tail at the rear. In retrospect one can only wonder if at any time it crossed Mitchell's mind to similarly divide the windshield—but it may have been too soon to "bring back" the split windshield panes of GM cars used as late as 1952!

Mitchell had made a step in the divided rear window direction on the Oldsmobiles for 1957 where he had extended two dividers through the glass area, splitting the window into three portions. "Twin-strut" styl-

1. Don't let anyone catch you doing this now, but by mid-1963 the "in" customizing trick was to remove the rear window divider, bond the severed juncture, and install a one-piece molded Plexiglass window and trim kit.

2. A first-time, $322.80 option for 1963 was the knock-off wheels built for Chevrolet by Kelsey-Hayes. Again available now due to demand and made from the same dies, figure on spending $895.00 per set which included four each of the wheels, spinners, center caps, adaptors and knock-off cones. Pictured is an original set on a well-preserved coupe, but with non-original tires.

ing, it was termed, and these were not simple chrome molding dividers, but continuations of character lines pressed into the roof body metal. A single, wide central divider, however, would block at least partially clear rearward vision. One of the early coupe mockups had a periscope mirror protruding through the roof above the center of the windshield, but (perhaps fortunately) this wasn't brought forward to final production.

Duntov, somewhat out of his element in the styling area, argued that thus-hampered rear vision would be dangerous and might also dissuade safety-conscious drivers from buying. He insisted that a car, especially one of high-performance persuasion and which would certainly see popular use in competitive events, should be designed to be seen out of. Mitchell's stand was that a car is designed to be looked at, and it is said he even threatened to cancel the entire Sting Ray project if his divided window was not brought to reality!

The window divider did in fact meet with some critiquing upon the Corvette's unveiling, but it turned out that visibility wasn't really all that bad from the driver's seat through the conventional interior rear view mirror. A single driver's side exterior mirror was required equipment anyhow, and most buyers would opt for a similar rightside mirror in the interests of symmetry if nothing else. Coupe owners—myself included—were so satisfied with their brashly beautiful and potent Sting Ray coupes that they would likely have put up with any inconvenience that Mitchell had seen fit to produce.

With the benefit of hindsight, and with original split-window '63's demanding such high prices today, it's almost a sacrilege to report that during the first model year, that many coupes were customized by the removal of the window divider and the replacement of the rear glass with a single piece of formed Plexiglas. Later, after introduction of the one-piece-window '64's, factory glass became available over agency parts counters and was fitted to numberless '63 coupes to the (present) dismay of all and sundry.

EYE OPENERS

The retractable headlights were an engineering challenge and several systems were tried before a final arrangement was selected for production. An early Sting Ray TV commercial showed a Sebring Silver coupe winking at a passing girl with one of its eyelids rapidly opening and closing. But this was photographed at slow camera speed and, when projected at the normal rate, made the repeated sequence very fast indeed. Too the other set of headlights had been rendered inoperative for the commercial, since in real life both sets of light covers operated independently but simultaneously.

Chevrolet correctly predicted later service problems with the twin electric motors and related Rube Goldberg-style gears and linkage, by noting in the Sting Ray owner's manual that the system could be overridden in case of failure. They neglected to point out, however, that one must have double-jointed arms and eyes on fingertips to find, then twist, the over-rider wheel. Neither does the manual describe the number of turns required to fully open or close the doors; the overrider turns at motor speed (converted Delco electric window motors!) and there is severe gear reduction between it and the lamp housings. The motor assembly, incidentally, is still available at $119.00.

The mere prescence of an early '63 Sting Ray was enough to attract a small crowd of curious onlookers, and for several months after receiving my coupe it never failed to provoke questions and comments whenever it was parked. Bystander comments were many and varied, but universally asked was the car's price and pleas rendered to operate the headlights.

BY THE NUMBERS

Opinions were divided within both Chevrolet in general and Corvette in particular, and between various car-admiring factions, over whether the convertible or fastback Sting Rays would enjoy the larger share of sales. Buyer's optional equipment choices were also unknown, especially for items like air conditioning and power brakes which were not previously offered, and factory-level procurers of such accessories were hard-pressed to second-guess future Vette buyers in their stock-piling of engines, transmissions, power assists, and so forth. Hesitancy in amassing large quantities of the specialty equipment may have helped delay full production at St. Louis. In the end, though, production of convertibles and coupes was astonishingly close, with over 49 percent of production devoted to coupes and the balance to convertibles. The model year saw both Sting Rays built to a 21,500-plus number, a figure some 30 percent greater than Corvette's best previous year when, in 1962, 14,500 had been manufactured.

Speaking of production numbers, it is interesting now to reflect on how some of the '63 factory options faired with buyers. When the results were in, the 4-speed manual transmission (in place of the base 3-speed manual) found a home in 17,950 Sting Rays. The detachable hardtop for the convertible went on slightly more than half of this style's total. Power brakes were ordered on 3,330 of the '63's, but power windows went on 3,740 cars. Air conditioning, a first-time Corvette choice, went into a scant 278 cars— and one of these is definitely a rarity today.

Summed up, the Sting Ray was a smashing success by Corvette standards; it handled, it performed, it looked absolutely great and— most importantly to GM, and to Mitchell's glee—it sold well. But this was only a harbinger of coming events. Even at the prodigious outpouring (for a "limited production" sports car) of the aforementioned 21,500 copies, this number was outstripped by every following model year to the present with the sole exception of 1970 when sales dipped to 17,300 units.

THE RACING FRONT

On the racing scene, the new Sting Rays were more slippery looking than their forerunners, thanks to Mitchell's design which was put to test, in scaled-down form, in the Cal Tech wind tunnel. But of major concern here were the forces that might cause the body configuration to lift at speed, rather than an all-out attempt to lessen frontal resistance. So while aerodynamics may have played only a small role in the Sting Ray's track successes, the chassis layout played a major one. The club and quasi-serious racers did well enough on their Sunday afternoon sportings, but serious although behind-the-scenes activity was taking place within Chevrolet's walls and on which

THE STING RAY EMERGES

Another first-time offering for the Corvette was air conditioning, but few Sting Ray buyers took the option. Only 278 so-equipped Corvettes were sold in the model year, but many buyers later had a dealer install the system. But only the factory-equipped cars included this rare decal.

This is the first "split-window" Corvette Sting Ray to reach the West Coast, and it's a safe bet that it was one of the first to fall prey to custom restyling. Owner Spence Murray took delivery prior to the model's public unveiling, turned it over to George Barris by January 1963. Barris repainted it in a rich metallic blue, adding a silver stripe to the hood blister and back over the top to visually enhance the rear window divider. Other touches were chromed, expanded metal panels in the hood troughs to replace the simulated aluminum air scoops, and matching insert panels behind the front wheels and in the top behind the doors. Trend-setting were the added taillights, a practice copied far and wide by Sting Ray owners at the time.

driver/writer John Christy expounds as follows:

"Anno Domini 1963 had been a bad year for Corvettes. Not to put too fine a point on it, America's premier sports cars had been snake-bit wherever they appeared, on the track and on the street. Sting Ray drivers had taken to skulking down side streets whenever they saw a Cobra coming down the pike, and Corvette racers were clamoring for Carrol Shelby's "snakes" to be reclassified—preferably into oblivion. It was a situation not to be endured. The Corvette, at least in its post-1957 persuasions, had taken America's highperformance enthusiasts by storm, knocking off exotic imports in races and bringing true road performance within reach of the common man. And now here was this—this hybrid —this Anglo snake-in-the-grass blowing them all into the weeds.

"What made the situation all the worse was that there was, apparently, no hope for a fix from Chevrolet. General Motors had clamped the lid on overt racing activities— and therefore supersonic Corvette development—in January 1963. It was especially ironic in that Shelby, and Ford, had served notice the previous fall at the Los Angeles Times Grand Prix and at the Bahamas International Speed Week that they had a real Corvette-killer underway.

"Most ironic of all, however, was the well-hidden fact that someone at Chevrolet had indeed taken notice of Shelby's early warning. Very quietly, an "Operation Mongoose" had been set up near Detroit. The plan was to build 1000 very special Sting Rays, 1200 lb lighter than normal, with enough handling and stopping ability to match the Cobra and enough power potential to blast the snakes off the track wherever they might appear. Some say the plan called for only 100 examples, but at this removed point in time the figure of 1000 seems more realistic, if you consider the number of dealers who might have been interested. In any event, the racing ban brought the project to a reluctant halt.

"But the halting of the operation didn't occur before five prototypes, three coupes and two roadsters had been built. The existence of these cars—or at least some cars— wasn't a total secret, but it wasn't exactly broadcast from the rooftops either. Through the arcane machinations that govern such occurences when factories and enthusiasts cross paths, two of the coupes were loaned, leased—in any event, released—one to Union Oil executive and Corvette enthusiast Grady Davis and another to Chevrolet Dealer Dick Doane. The racing dentist, Dr. Dick Thompson, drove the Davis car throughout the spring and summer of 1963, ending up with a class and overall win at Watkins Glen. Doane also raced the second car to some extent. Strangely enough, little notice was taken, at least in the public prints, of these cars—possibly because nobody was at any pains to call attention to them. To most of those who did notice them, they were probably simply put down as two more modified Stingrays.

"As that disastrous—for the 'glass Corvettes—summer of 1963 wore on and the snake plague continued unabated, somebody who had more than a modicum of decision—making power at Chevrolet decided the situation was not to be borne. Somehow, some way, Chevrolet had to get in at least one inning. But how? They had the potential weapons in the GS Corvettes, but they weren't production cars, so a head-to-head confrontation in some major SCCA event was out of the question. As non-

Prior to the grand assault on the sports-racing world at Nassau in December 1963, two of the GS coupes did well under the capable guidance of Dick Doane and Dick Thompson. They were white (until they appeared at Nassau in the Mecom team's Cadillac light metallic blue) and had sprouted scoops and trimmed front fenders to solve engine compartment heat. The acceleration of the-then fitted 377 cu.-in., 480 hp engine was described as "stupendous" with some thanks to the GS's weight, 1995 lbs. "wet."

production cars they would be up against the new breed of mid-engine sports racers that were just coming in, forerunners of the ferocious Can-Am machines, not against the Cobras.

"Then someone had a brilliant idea: Bahamas Speed Week, otherwise known simply as Nassau Race Week. This was a unique event, traditionally the season closer, as much a weeklong party as it was a race. It had two important things going for it from Chevrolet's standpoint. Number one, Speed Week was a promotion backed by the Nassau Tourist Bureau and the local hotels. As such it drew press and media coverage far out of proportion to its importance in the normal scheme of racing events. This coverage was not at all hindered by the fact that members of the enthusiast press were given free hotel accommodations, thus reducing the expense account bite. Number two, and just as important, was that the rules were set up by the dictatorial and flamboyant promoter, Capt. Sherman F. "Red" Crise. To Red Crise the show was everything, and rules were made to be bent in the interest of show.

CONFRONTATION

"If Captain Crise was convinced that he had a show in a confrontation between Cobra and Corvette, then by any means there would be just such a confrontation. This bit of hybridization of the competition classes was easy in that, during the two major events, the Governor's Trophy and the Nassau Cup, everything still in one piece ran together. With a 4.5-mile course, starting fields of 60 cars were not unusual. There was no way a head-to-head confrontation between the GS Corvette and the Cobra could be avoided, and right under the noses of a very heavy contingent of the world's print and electronic media.

"The two loaned-out coupes were called in and, with the third coupe, were quietly prepared up to the point of bending the letter of the very loose Nassau Grand Touring class rules. Totally bereft of anything resembling interior trim, they were fitted with engines poked and stroked to 377cid.. Fitted with four dual sidedraft 58mm Weber carburetors and all the necessary interior equipment, these engines cranked out 480 hp somewhere in the neighborhood of 6500 rpm and produced 500 lb.-ft. of torque at 4000 revs. Ugly but effective wheel flares were added to cover the 9-in.-wide Halibrand mag wheels and oversize Goodyear stock car tires. Memory fails to serve and history recordeth not the exact size of these tires, but they were, in fact, too big in terms of diameter. The car had been engineered for low profile tires which were not available at that time, and to get the size of footprint to handle the power they had to use what they could get.

"There was one question remaining. Who would be the entrant of record? Chevrolet couldn't, by edict from GM management, so a plausible private party had to be found.

"Enter stage left one John Mecom. Second-generation Texas oilman, now owner of the New Orleans Saints pro football team, Mecom was then into road racing, having fielded a small entry at Nassau the year before Exactly how the powers at Chevrolet and Mecom got together remains something of a mystery to this day. However it happened, Mecom had to be the ideal entrant of record. Enormously wealthy (Need a spare engine? Call out the company 707), he was entirely plausible as a major racing team owner. He hadn't made enough of a splash the year before to even be noticed in the public prints, but this year was to be different.

"When the transport ship docked in Nassau, it off-loaded five light metallic blue cars—three GS Corvettes, the LeMans Lola coupe and the ex-Reventlow mid-engine Scarab, all stuffed full of hairy power by Chevrolet. The listed drivers for all this ferocity were Jim Hall, Dick Thompson, Roger Penske, Augie Pabst—and A.J. Foyt. By the time the week was out, the combination had taken everything but the ladies' race and the Grand Prix of Volkswagens.

"On the very first day, notice was served that this was one occasion when the Chevy-powered machinery was not going to sit and watch the Fords go by. In the first 5-lap opener, two Corvettes and the Lola just flat ran off and hid from everything else including the previously invincible Cobras. At the finish it was Pabst in the Lola, followed by Thompson and Hall in the GS Corvettes. In the following 100-mile Tourist Trophy they again proved their point—to a point. The two Grand Sports and the Lola again took off in the lead, but this time the brutal twist of the 480 horses in the Vettes put so much pressure on the rearend gears that they lasted 9 and 15 laps, respectively, before they let go.

"On the following Friday, some 58 cars lined up for the Governor's Trophy. When the flag dropped, Penske in GS #003 charged off in the lead. For the rest of the 25-lap event he was passed only by Foyt in the Scarab, Hall in the Chaparral and Pedro Rodriquez in a Ferrari 250LM. During the race, Hall made an agricultural excursion into the bush, and at the end Penske was 3rd overall and 1st in Prototype class. Fourth and 2nd in class was Pabst in another GS. Dick Thompson, in the third GS, was not running at the finish but had piled up enough laps to be credited with 6th overall and 3rd in class. The Cobras weren't even close enough for a mention in dispatches.

"In the Sunday finale, the Nassau Trophy (62 starters and 56 horrible laps), two of the GS Corvettes did it again with #004 taking 4th overall and 1st in class and #003 coming in 8th overall and 3rd in class. The performance is all the more remarkable when it is recalled that a sizeable proportion of that 62-car field was of the mid-engine Group 7 or Sports Racing category and a third of the field was powered by hopped-up V-8 Detroitware of up

THE STING RAY EMERGES

to 427cid. They might have done even better, had it not been for the tire choice. The large-diameter tires held the front ends high, and the resultant lift and piling up of high-pressure air in the hood compartments caused the hoods to pop open on several occasions. Had it not been for the pit stops to resecure the hoods....

"The rash of publicity that followed this performance told the world that Chevrolet could do the deed when it wanted to. It also had the undesirable effect of causing management to reiterate its ban on overt competition. The GS Corvettes never ran again as a single team. GS #003 was retained by Mecom for two years. It was entered at Sebring in 1964 with Foyt and John Cannon as drivers, running as high as 8th during the first five hours until a wheel came adrift. It finished 23rd. It showed up again in the '64 Nassau race but with indifferent results. GS #005 had gone to Jim Hall, and it too showed up at Nassau in '64, where Penske again took a class win with it.

INSIDE AGS

"Just what were these GS Corvettes and how did they differ from the regular Sting Rays of the time? First of all, as mentioned, they were LIGHT. The frame is made up of large-diameter tubing instead of pressed channels. The body thickness is about half that of the normal Sting Rays, and aluminum is used wherever possible to replace the steel components of the standard car. The passenger compartment framing, the underdash cross brace, window channels, seat bracing, door handles, brackets and hood brace are all light alloy. In the original car the dash panel was vacuum-formed plastic instead of the standard heavy potmetal. Light alloy chassis parts include the differential housing, steering box and portions of the brake calipers. The total body weight, including windows, doors and hood, is 375 lb. All windows except the windshield are Plexiglas. The wheels are by Halibrand, 15 x 8.5 in front and 15 x 9.5 in the rear, and made of magnesium. It is estimated that, with the new aluminum engine, restored GS #003 will tip the scales at about 1995 lb, wet!

"Add to this mix the 480 hp and 500 lb-ft of torque, and it is clear how the GS could show a clean pair of heels to the 289cid Cobras of the time. It all makes you wonder what would have happened if Chevrolet had gone ahead with plans to build and market 1000 of these GS Corvettes. Would it have spurred Shelby—and Ford—to even greater efforts, or would it have discouraged them to the point of packing it in? Of a certainty, the GS must have pushed the development of the 427 Cobra more quickly than it might have been. It undoubtedly also hurried the development of the Cobra Daytona coupes, which went on to take a World Championship for the U.S., and that can't be all bad.

"Whatever the conjectures, the fact remains that the GS Corvettes were the fastest front-engine GT-type cars of their day. But then, Chevrolet wasn't racing, was it?"

Of the five Corvette Grand Sports—three coupes and two roadsters—that lived through the GM ban on racing activities, only this one exists in restored form. It is GS #003, the last of the coupes to be built, and presently belongs to Bob Paterson. John Mecom had retained it for two years after the '63 Nassau bash, running at Sebring in '64 with A.J. Foyt and John Cannon driving. These two managed an 8th spot overall in the first five hours, but finished a dismal 23rd after a wheel left the fray. Corvette GS #003 returned to Nassau in '64—as did GS #005 under Roger Penske—but made a poor showing. It went through various owners after that (including one who street-drove it,) but finally came to Paterson who had Bob Waldschmidt of Automotive Enterprises in San Carlos, CA, restore it as nearly as possible to its true '63 Nassau self. Its powerplant at the time of photography is a blueprinted 327, but Paterson, with the help of Smokey Yunich, has unearthed an aluminum small-block (similar to the Jim Hall Chaparral powerplants) and is rebuilding it to original Nassau specs but with 45mm Webers instead of the 58mm originals.

The hunched rear fender duct feeds air to the brakes; the holes poked between the taillights let air escape from under the floorless trunk area. Recessed doorhandles keep these necessities out of the slipstream, as does the inset quick-fill fuel inlet buried above the right rear fender. The fender flares, as author Christy points out, were ugly but effective.

THE SECOND GENERATION

PHOTOS BY GM PHOTOGRAPHIC, COURTESY JOHN AMGWERT

PHOTO BY GREG SHARP

The Corvette was sprung on an unsuspecting public when, for the 1963 model year, automotive design in general was largely at a standstill. Styling Chief Bill Mitchell's bombshell included not one, but two versions of the Sting Ray; convertible model 867, and coupe model 837, both fitted to a 98-in. wheelbase, 4 ins. shorter than the previous "big" cars. GM Photographic did its usual justice to them and they deserved it; knife-edge styling, retractable headlights, fastback styling for the coupe with its controversial and, as it turned out, short-lived divided rear window. Pilot-run cars, which these two were, nearly always include all available options, and both of these handsome machines carried the fuel-injected 360 hp 327 V-8, option L84. The sensation, though, for an American production car was described without fanfare on dealer's order forms as: "Wheels: Five cast aluminum 15X6L quick knock-off type". Made to specifications set down by Chevrolet by Kelsey-Hayes, they listed at the time for $322.80. Current demand has put them back in production for nearly four times this amount.

Zora Arkus-Duntov had done his homework as well as Mitchell had, and under the sleek skins lurked independent rear suspension which would bring at last to sports car purists handling and roadability previously unknown. Zora took exception to Mitchell's handiwork; he felt the rear window divider would hamper driver vision (he ordered one-piece rear windows in his three GS racing coupes without Mitchell's knowledge), and he worried over wind tunnel reports that the cars tended to lift at speeds over 100 mph.

The Rochester f.i.-equipped engine illustrated reveals in part some of the tender, loving care being lavished on early Sting Rays today.

THE SECOND GENERATION

PHOTO BY ERIC RICKMAN

Most Corvette restorers haunt the specialty shops seeking the hard-to-find components they need to put the final touches on their labors of love. But Chris Wickersham did it the other way around. He opened his own store by stocking it with the left-over bits and pieces he had after dismantling several '63's in order to restore this one convertible. The photo reflects the lengths to which serious restorers go; infinite care (and expense) in returning the chassis to like-new condition, care in using only those optional extras with which the car was originally equipped, duplicating the plating and painting of every individual part after stripping the chassis to its bare rails, and making sure the frame, engine, transmission, distributor and fuel injector body numbers all agree. When incorrect number coding is found by zealous show judges, the car may be docked points as a "counterfeit".

Chris owns five more Corvettes of '63-'65 vintage including another like this one but in low-mileage, original form and valued at $20,000. To show how early Sting Ray values have escalated in the past few years, Chris got this one in running shape for $1250 in 1973 but will have another $20,000 car when restoration is done.

PHOTO BY ERIC RICKMAN

A '64 Sting Ray undergoes testing at GM's Milford Proving Grounds and shows up some of the subtle exterior refinements. The chrome-edged windshield posts used in '63 were gone, and the painted posts helped integrate the coupe's roof and body sides. The twin hood depressions were without the earlier aluminum inserts and new rocker moldings were more boldly striped. To the dismay of a few but the delight of all others, the rear window divider had disappeared.

Underpinnings were upgraded, too. New variable-rate front and rear springs, recalibrated shock absorbers, new rear engine mount and rubber body mounts, plus additional sound deadening and cushioning materials, all helped the second-year Sting Ray ride, handle, and feel better.

RIGHT: Refinement inside and out was the byword for 1964. More easily read were the dark-hued instrument dials and a simulated-wood rimmed steering wheel was standard. The central floor console was redesigned. A rarity today is this convertible with Four-Season air conditioning. It also carries fuel injection, 4-speed transmission, and power steering, brakes and windows. options not often found together.

BELOW: A pair of convertibles pose grandly in Fountain Valley, CA, home of Corvette enthusiast Jerry Cogswell. His is the original, black '63, and it's a "fuelie" with 4-speed and those ever-lovin' knock-off wheels. The red '64, and whose interior is also shown here, was restored by Jerry. It's owned by Helen Bode, Olathe, KS who once saw Jerry's '63 and "wanted one like it". He told her to find one and he'd restore it for her. She did, so he had to; and with outstanding results.

PHOTOS BY GREG SHARP

RIGHT: Probably not really typical of original '64 coupes today, Jim Richey's has had only one repainting of its Riverside Red exterior yet it's a consistent trophy-taker in concours displays. Now pushing toward 100,000 miles on its clock, the fastback has never had its 300 hp version of the 327 V-8 "opened up". Still carrying all of its as-built options, it sports air conditioning, power windows, steering and brakes, automatic trans and those ever-lovin' knock-off K-H wheels. Jim is a member of the So Cal Vettes and lives in Costa Mesa. Like all good dyed-in-the-wool Vette buffs, he obviously spends much of his spare time keeping the prize in show condition, and he shows and tours regularly with his fellow members.

PHOTO BY JON C. JAY

THE SECOND GENERATION

PHOTO BY GM PHOTOGRAPHIC

Basic styling trends of the coming generation of Corvettes was begun by Bill Mitchell as soon as the '63 Sting Rays hit the dealerships, a stylist's interest in a project understandably waning by announcement time after the several years' close involvement to bring drawing board ideas to reality. Thus did he "do" the Mako Shark II using the talents of designer Larry Shinoda and the Styling staff. The name came from the original Shark, a flambouyant show piece of 1961 and which was later dubbed Mako Shark after its color scheme that blended gradually from its under-belly white to blue. Mitchell wanted the same treatment on the new car, so the II appellation was given. Built first as a mock-up but on a "real" '63 chassis, the non-running creation was completed early in '65 and put on the show circuit, while a fully operable twin (with certain refinements) was being constructed. It was finished by October of that year when this photo was taken at Milford.

PHOTOS BY BOB NAGY

Despite their present age, Corvettes of the '63-'67 era continue to wage road-racing wars quite successfully. Eric Weinrich's '65 coupe runs widely flared fenders but otherwise runs the standard body. It competes in SCCA's B Production class in West Coast regional events. Bob Ryan's convertible competes in A Production class, running sans full windshield and thoroughly gutted. The A Corvettes run up to 454 cu.-in. engines, the B's 350's.

THE BETWEEN YEARS: '64 TO '67

IT WAS A PERIOD OF REFINEMENT ON THE OUTSIDE, AND GREAT STRIDES UNDERNEATH

It was spring of '65 and rain darkened West Philadelphia. We parked the Ford to search the used car lot for Corvettes. It was our fifth Chevy dealership that day and in our search for a clean used Sting Ray, John and I had found nothing. But with this dealer our luck might change. This was Roger Penske's Chevrolet and if anyone had a good used Vette, it would likely be Racer Roger.

Sure enough, back in a corner were two Sting Rays—a coupe and a convertible. They were both '64's and appeared well kept. Following a closer inspection, we agreed that the coupe had received better care than the convertible and besides, John preferred the coupe's fastback design. Undoubtedly, we were being watched, but to the salesmen, we were just two more Walter Mittys—lots of dreams and no cash. With hands cupping our eyes, we studied the interior through the window. Complimenting the silver blue of the paint was a dark blue interior, one of the six colors available that year in vinyl or leather. Matching vinyl covered the hoods of the instrument panel and the upper halves of each door panel. Dark blue carpeting softened the floor and extended halfway up the door panels. The luggage area was also carpeted.

I had been hooked ever since I saw my first Corvette in '53, but in the past month we had both become students of the Sting Ray, searching magazines and reading every Sting Ray article we could find. We spoke to Corvette owners and looked at a number of used models and now we were ready to test our new-found knowledge. We examined the car for inaccuracies, flaws and evidence of a possible counterfeit.

Apparently unhappy with the split window, some nefarious '63 owners had removed the divider and in its place installed one-piece glass. A few other cosmetic changes and perhaps a new i.d. tag beneath the dash could convince most people that the car really was a '64. Ironically, just 10 years later, the opposite would occur—'64's would be changed into '63's.

The coupe at Penske's appeared to be a genuine '64. Aside from the one-piece rear window, GM had incorporated a number of new features into the '64 and on this particular car, we spotted nearly all of them. On the '64 models the simulated air intakes in the hood were eliminated, although their indentations remained, and real louvers were added to the roof pillars in the coupe; however, only those on the driver's side were functional. For '64, the rocker panels beneath the doors included fewer ribs than the '63 panel and the area between the ribs was painted black. Another change was the fuel filler cap which was more ornate with the addition of concentric circles around the crossed flags; The '64 wheel covers were simpler in design and now resembled the smooth, slightly conical look of the popular '57 Plymouth cover.

Inside, plastic trim was upgraded for '64. Color-keyed to the interior, the plastic steering wheel rim had

1. To the dismay of some (Styling boss Bill Mitchell for one) but the glee of others (Engineer Duntov for another) the divided rear window was replaced in '64 with one-piece glass. Other subtle refinements included louver panels over operable air vents (on the left side).

2. The convertible enjoyed an increase of sales percentage (split about 50/50 with the coupe for '63, but up to 63 percent in '64). Like the coupe, the convert carried more boldly striped rocker moldings and the hood indents lost their aluminum "intake" panels.

THE BETWEEN YEARS: '64 TO '67

been replaced with a wood-grained plastic and it had the convincing look and feel of varnished walnut. Its attractive 3-spoke hub had been retained. The gearshift knob, inside door knob and turn signal stalk were now chromed steel, while they had been plastic in '63. Criticized for its glare in '63, the aluminum backing of the instrument faces was changed to flat black for easier reading.

HE BUYS THE '64 COUPE

We checked out the chassis for less apparent refinements; and found that the coils on the front springs and the leaves on the rear spring were of varying thicknesses. These were the variable rate springs we had read about. They were designed to flex according to demand; that is, the larger the bump, the more the spring flexed to absorb it. This reduced ride harshness without sacrificing handling quality. The double-acting shocks were recalibrated for a smoother ride, bringing flatter cornering and less body roll in the turns. When subjected to hard use, shocks overheat and cause the hydraulic fluid to bubble and efficiency to drop. So for '64, a small bag of freon gas was placed in the fluid reservoir of each shock. The freon absorbs the heat, foaming is therefore prevented and shock efficiency remains high.

Cockpit noise was a problem for '63, so for '64 engineers reduced the sound level with more insulation, new rubber body mounts, and added a more flexible transmission mount, a new shift lever boot and, to reduce the buzzing of the shift linkage, more rubber linkage bushings. Exhaust noise was decreased by isolating the pipes from the frame.

A 3-speed all-synchro trans was the standard '64 box, but only 3.2% of the cars produced were so-equipped. The stocky shift lever of the optional 4-speed was a reminder that this car had Corvette's new Muncie unit. Replacing the Borg-Warner T-10 in mid-'63, the Muncie was a close copy but carried stronger synchronizers and wider gears for smooth, grind-free shifting. Two ratio choices were offered. Available only with the lower-hp 327's, the wide-ratio box was geared with 2.56:1 low, 1.91:1 second, 1.48:1 third and 1:1 fourth. Sold only with the 365-hp engine and the fuel-injected (375 hp) block, was the close-ratio model: 2.20:1 first, 1.64:1 second, 1.28:1 third and 1:1 fourth. Regarding ratios, both Muncies were very similar to the two T-10's of '63 and like them the Muncies featured a lockout reverse trigger.

The tach was redlined at 5000 rpm. We knew that the 365 engine and the fuelie redlined about six grand. Therefore, unless the engine had been changed or modified, this was a 250- or 300-hp block.

Having nearly completed our 20-minute perusal, we noted a salesman strolling over to investigate. John expressed his interest in the car and must have beamed good vibes or a deceivingly thick wallet because the salesman offered him a test drive. The engine came to life with a turn of the key. Its exhaust note was deep and mellow, the idle smooth and the absence of oil smoke from a cold exhaust indicated a healthy engine free of valve guide problems. The coupe turned right on Chestnut Street, fish-tailed slightly on the wet pavement and disappeared in traffic. Ten minutes later, the car returned and both men were smiling. Obviously happy with the experience, John had fallen in love with that plastic Chevy. We followed the salesman to his office and after the traditional haggling, they agreed on a price—$2600 and the '61 Ford. Just a year earlier, the basic price for a '64 coupe was $4364. Adding $188 for the 4-speed, $86 for power steering and $171 for its AM/FM radio, the Penske car sold new for $4815. Allowing $1500 trade in on the Ford, John was paying $4100 for the used car—a high price but to be expected when considering its high demand and short supply. A total of 22,229 Corvettes were produced in '64 with 37.3% (8304) of those coupes—and so far very

1. The faithful old Borg-Warner T-10 4-speed had disappeared and for '64 Chevy was touting its all-new trans built in Muncie, IN, and thus acquired that city's name. It had stronger synchronizers than the old T-10, and wider gears. Both close- and wide-ratio gearsets were available.

2. Roadability, handling and performance were the Sting Rays' forte; for the shear fun of driving, a road like this and a Corvette were a matched pair.

3. While a styling mockup of the first Sting Ray coupe included an openable deck lid, cost prohibited its use in production and Sting Ray owners had to settle for through-the-doors loading.

4. You could throw a little fear into the life of a Cobra driver if your Corvette sported this now-famous emblem on its front fenders.

few of those had materialized on the used car lots. An oil change, lube and a wax job accompanied some mechanical repairs, and by Saturday, the car was ready for the 80-mile run to its new home.

It was my first ride in a Corvette. We swung the seat-backs forward and loaded our luggage. At 10.5 cu. ft., the area was roomy but inconvenient when loading large, heavy bags through the door and over the seat. Furthermore, it offers no privacy for hiding large valuables. However, a section of the floor doubles as a cover for twin concealed compartments—an excellent place to hide small valuables. Extending from wall to wall, the compartment is deep and narrow and divided in two by the driveline hump. On the left were stowed the scissors, jack and lug wrench. The Sting Ray was yet a stranger and, as a used car, its reliability was unknown. We packed our tools. Topping the tank with good ol' Sunoco 260, we realized that this coupe had the standard 20-gal. tank and not the 36.5-gal. ($202) option.

At the city's crawling speed, air circulation was minimal, but it did afford the opportunity to try the Corvette's new air exhaust system. A pull cable under the dash opened a flap in the roof pillar and a twist of a special knob turned on a blower situated well behind the left rear wheelwell. Unfortunately, we weren't the only ones getting hot. From the hood drifted a puff of steam and the needle on the temperature gauge approached its peg. After taking the first available turn onto a side street we pulled to the curb. John shut down the engine and popped the hood. Apparently, the slow traffic and the warm air were too much for the cooling system. The aluminum radiator was quick to dissipate the heat and soon we could release the pressure at the expansion tank. Since the car had spewed considerable coolant from the overflow, we agreed that it should be replaced before going much further. A chemical coolant is essential to the life of an aluminum radiator; without it, the inside will quickly scale and clog the flow of water. Our smoke signals had done some good—two young ladies emerged with pitchers of water. Of course, we would have to wait for the engine to cool before adding the water—a perfect excuse to prolong the conversation. Thirty minutes and two phone numbers later, the radiator was full, traffic was lighter and we were on our way. The first miles were marked with traffic circles—notorious landmarks in South Jersey. Like the other traffic, we entered the circle at 50 mph but our quick steering was a big advantage when combined with the responsive 250-hp engine and efficient drum brakes.

On the 3-lane highway, the blue coupe maintained a smooth 70 mph. The standard 3.36:1 Positraction axle let the engine relax at a comfortable rpm. Inhaling through the small throttle bores (1.44-in. primaries and secondaries) of a Carter WCFB 4-bbl., the factory-stock Vette did surprisingly well on a tank of gas. Later calculations showed 20 mpg for the mostly highway trip. The small engine, the 4-speed and that standard 3.36:1 axle made a nice combination. The smallest of Corvette's 4-engine stable, the 250-hp 327 reaches its peak power at 4400 rpm. Hydraulic lifters, a mild cam, a single-point distributor with vacuum/centrifugal advance, a forged steel crank and dual exhaust combined to make this en-

The K-H-supplied knock-off aluminum wheels were carried forward into 1964. The RPO P48, for $322.80 list, brought you five of the 15X6L wheels, four adaptors, four 3-lug knock-off hubs, and a special knock-off hammer. (We wonder how the NOS market on the latter is doing!) Motor Trend magazine got a fuel-injected coupe for its annual road test and in between runs at Riverside Raceway they demonstrated the wheels. First, the hammer was stowed with the otherwise-standard jack and handle inside the rear left floor bin. The knock-off lug was rapped a few times, then spun off. The cast-aluminum alloy wheels made changing painless, and one-handed handling was easy. With wheel off, tester indicates venting holes that help keep brakes cool.

THE BETWEEN YEARS: '64 TO '67

gine a dependable unit with none of the frequent adjustments and problems of the larger, more sensitive engines. With a compression ratio of 10.5:1, the 250 shared the same pistons with the optional 300-horse block. The additional 50 hp came from a larger 4-bbl. and to handle the extra fuel, bigger intake valves and ports, a cast-iron intake manifold and larger exhaust manifolds. The cam, lifters and distributor remain the same. The Carter AFB (aluminum 4-bbl.) featured 1.56-in. primaries and 1.69-in. secondaries—more power but naturally, more fuel. Pleased with the base engine, John mentioned that anything bigger would have been impractical for his frequent commutes between school and home.

BIG ENGINE BEATS

However, the feel and the sound of 365 horses was hard to resist. The rough, loping idle and the subdued clatter of the solid lifters were constant reminders that this a sensitive piece of machinery; an engine that required attention and frequent adjustments. A jab of the throttle brought instant response with a raspy but beautiful exhaust note. The additional 65 horses are the result of a number of modifications: Intake valves were increased in size from 1.94 ins. to 2.02 ins. and exhaust from 1.50 ins. (300-hp engine) to 1.60 ins.; Holley's new 4-bbl. and to handle the fuel, a large 8-port aluminum intake manifold. The new 4-bbl. was very similar in function to the Carter AFB and they shared the same size primary throttle bores (1.56 ins.) but at 1.56 ins., the Holley secondaries were 0.13-in. smaller than the AFB. In appearance, they looked totally different. The secondaries on the Holley open by vacuum and close mechanically. The Holley offered new benefits of easier on-car adjustments. A power valve or a jet change could be performed quickly by simply removing the float bowls and the metering block while the AFB required considerably more disassembly. With an adjustment screw on the top of the bowls, float level could be adjusted with the car running—a 10-minute job. A new camshaft replaced the solid-lifter Duntov cam that had performed so well since '56. At 0.485-in., intake and exhaust, the new cam had a considerably higher valve lift than the 0.3937 intake and 0.3997-in. exhaust of its 340-hp brother of '63 and a longer duration too. To accommodate the additional lift, the tops of the pistons had to be machined for valve clearance. This created a dome effect which reduced the compression from 11.25:1 (1963 340 hp) to 11:1 for the '64 365-hp block. Protecting the valves was a pair of finned aluminum valve covers, while the valve covers on smaller engines were Chevy-orange steel. Protecting the crank was a larger oil pan, 5 qts. instead of 4. John and I had seen the bigger engines perform in street races and perform this one did. At 375 hp, the fuelie was the king of the hill (that is, next to the Cobra). Aside from the injection, the engine was the same as the 365-hp version but in many ways, the fuel injection was better than the carburetor.

At $538, the Ramjet fuel injection was an expensive option but a far more efficient method of distributing fuel. Fuel-injection nozzles are mounted in the cylinder head just above the intake ports. The system meters a precise amount of fuel through the nozzles and when mixed with air, a nearly perfect fuel/air mixture is delivered to each cyl-

1. The base '64 wheel covers carried the popular 3-eared spinner treatment, but the discs now sported radial slots for improved brake cooling.

2. MT's assistant technical editor, and a handler as well, wails through Turn 9 at Riverside after a run down the long backstretch at 134 mph. Repeated emergency stops from between 100 and 130 mph failed to produce dangerous brake fade, but as one of the staff observed, "They smelled terrible".

3. King-of-the-heap was the Rochester f.i.-equipped 375 hp 327 engine, RPO L-84, going out the dealer's door for $538 over the base 250 hp engine.

4. Motor Trend's '64 test model shows itself off. Knock-off wheels, 375 hp "fuelie" engine, 4-speed trans and 4.11:1 rear end made quite a car out of it and duly impressed the magazine's staff. They determined the best shift point was 6000 rpm and clocked 3.0, 4.4, and 5.6 secs. from dead stops to 30, 45 and 60 mph respectively. They managed the standing quarter in 14.2 secs. at an even 100 mph.

inder. Consequently, the driver enjoys instant and smooth throttle response at any speed and no fuel flooding or starvation when leaned into a corner. Unlike the carburetor, f.i. constantly corrects itself to humidity, temperature and altitude changes. One of the greatest benefits from Corvette's mechanical fuel injection is its fuel economy. Although producing 375 hp, the injected 327 achieved a 12-15 mpg average in city/highway driving and as much as 21 mpg highway. And of course, performance was unmatched. Tested by Motor Trend magazine (September, '64) a '64 injected coupe with standard 4.11:1 axle required only 5.6 secs. from 0-60 mph and completed the ¼-mile in 14.2 secs. @ 100 mph. MT's Bob McVay reported: "Acceleration in all speed ranges was, to say the least, fierce—of the smash-you-into-your-seat variety. The engine proved strong and willing and pulled strongly right up to 6700 rpm and beyond in every gear."

With the traffic circles behind and open road ahead, we relaxed and enjoyed the comfort of the car. The seats were low, making it difficult when parking to see beyond the peaked fenders but experience creates a sixth-sense feel and, in time, it would be no problem. Adjustable fore and aft, the bucket seats were sprung soft enough to be comfortable yet sufficiently firm for adequate support, with one exception: The seatback was straight and lacked padding for the lower back. After 50 miles a dull pain began, but as a six-footer, I found enough legroom for stretching and position changes. Two bolts at the base of the seatback permitted a range of adjustments for rake angle. The steering column was also adjustable through 3 ins. of fore and aft movement by merely loosening an underhood collar on the column, adjusting the wheel and tightening the bolt. New for '64, improvements in body mounts, insulation and ex-

With the ever-increasing sales that the Corvette was enjoying, and hence growing numbers of the plastic car encountering body damage, Chevrolet was beginning to feel that the high costs of fiberglass repair might scare potential buyers off. They came up with a solution during 1965; a set of 17 patterned body molds that covered all body areas except the doors and hood. Where severe damage demanded replacement of that immediate area, instead of simple patching, that area was cut off the car completely and the appropriate mold was clamped in place (or abutting molds could be bolted together if the damage spanned two or more of them). A scribe was used to mark the outline of the cut-away portion on the under side of the mold, then the mold is removed. A new section is built up in that mold to the designated outline, using fiberglass, resins and a hardening agent. When cured, the part was taken from the mold and glued into place using a flange bonded to the existing body section. The seams are then filled and the repaired area sanded and, finally, painted. Theoretically, then, an entire body (minus doors and hood) could be layed-up in one piece, since a unit female mold was provided by bolting all the 17 molds together. Has anyone seen a full set at a swap meet?

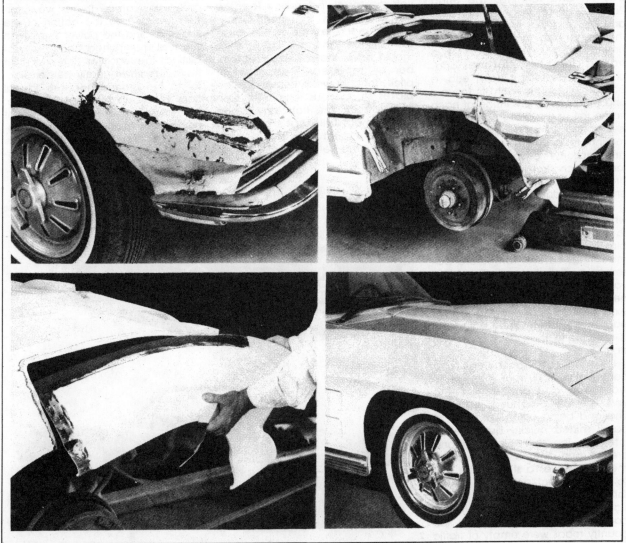

THE BETWEEN YEARS: '64 TO '67

haust noise were successful because the coupe was very quiet. As the only radio available, the optional pushbutton AM/FM was of superior quality and provided sensitive, static-free reception, excellent tone with no drifting from station to station.

IN THE DRIVER'S SEAT

After a weekend of waxing and detailing the car, lying on the beach and nightly cruising, we headed back to the city and my first personal experience behind a Corvette wheel. It was a different world in the driver's seat. Flooring it from a standing stop, I was impressed with the car's instant response and, particularly, its minimal wheelspin. The rear of the car squatted on its suspension, the front fenders rose and that coupe was off and running. The Sting Ray's excellent weight distribution and i.r.s. worked together to get the weight and torque to the rear tires. The 3-link i.r.s. features two trailing arms, two strut rods and, universally jointed at each end, a pair of axle shafts. With its forward end mounted to the inside of each frame rail and running parallel to it is a control arm. Each hub spindle mounts to a control arm and the arm absorbs drive and torque forces. Each strut rod pivots from a bracket mounted under the differential. The rods extend outward to a mount below each hub assembly where they pivot on a strut rod shaft. One end of the shaft doubles as the shock absorber's lower mounting. Lateral forces are absorbed by the strut rods. The 9-leaf variable-rate spring is bolted beneath the differential carrier and extends transversely to the rear of each control arm where each end is secured to the arm with a long bolt.

I wound out in first gear to a conservative 4000 rpm (after all, it wasn't my car) and shifted into 2nd. The car continued pulling in 2nd and being unfamiliar with the car, I shifted quickly but carefully into 3rd. The shift was quiet and precise and, with a slight push to the right, it was a straight-line shift to third—perfect action for speed-shifting. With the approaching stop sign, I backed off at 80 mph and came to a straight-line stop. The country roads were empty of traffic

so we experimented with several more high-speed runs. We calculated 0-60-mph time as approximately 8 secs.—not bad for 250 hp. Repeated braking created some fade and a mild pull to one side so we stopped our tests to let the drums cool. The Z-06 brake option was offered again for '64 but only in a package deal with fuel injection, 4-speed and Positraction. It consists of segmented metallic brake linings for better heat dissipation—12 segments in front, 10 for the rear. The 11-in. drums were finned for cooling and the front brakes received outside air through special ducts on the backing plates. Integral fans circulated the air to the brakes. Unlike conventional systems, these brakes self-adjusted in reverse and when going forward. When subjected to prolonged hard service, drums get hot and expand away from the shoes resulting in improper contact and brake fade. The forward adjusting mechanism compensates for drum expansion by moving the shoes closer to the drum. Included in the package was a dual-reservoir master cylinder; one feeding the front brakes, the other supplying the rear. The road testers loved the option and at $629, they should have. Motor Trend (September '64) reported: "Regardless of price, though, we're impressed. The brakes stopped us time after time in a straight line from any speed and without fail. That's about all you can expect from any braking system." We felt that for all-out

competition, the price is justified but for the 250- or 300-hp commuter, the stock asbestos linings worked just fine. However, for frequent mountain driving or towing, optional metallic linings might be a good investment. Although not in the $629 class, the sintered linings were said to perform much better than the asbestos. Metallic linings work best when they're hot; the more frequent their use, the better they stop. Fade is minimal but pedal pressure is greater so power assist ($43) is a recommended companion.

The back roads were the perfect excuse to get acquainted with the Sting Ray's suspension. We found a stretch of road with a varied sequence of curves, and, taking turns, we experimented with the "course." Starting off slowly, we acquired a feel for the car and gradually increased speed. At 80 mph, the car remained predictable and cornering was flat and stable. As the car powered into the curves, wheel deflection increased but as it did, the progressive-rate springs provided greater stiffness for minimal lean. Lateral forces were further reduced by the action of the strut rods and the power transfer to the rear maintained tractional stability at the rear tires. The tires must have remained nearly perpendicular to the ground because squeal from tire scrub was not excessive. At that speed, steering felt almost neutral but more throttle on a decreasing radius turn would have probably induced enough under-

steer for a loss of control. The power steering action was fast with an overall ratio of 17.6:1 and from 3.4 turns to 2.9 by simply moving the tie rod ends from the rear hole in the steering arm to the forward hole. Incompatible with the sophisticated suspension, the factory tires limited the Sting Ray's potential handling abilities. Furthermore, limited the Sting Ray's potential handling abilities. Furthermore, when combined with the Positraction differential, the tires were unpredictable on wet or icy pavement. The differential is designed to send the torque to the tire with the most traction but when both tires are on a slippery surface, the torque transfers from wheel to wheel causing the car to fishtail. Consequently, the driver must back off the throttle and tread very carefully to dry pavement. Although, when one tire is stuck in snow or mud and the other rests on a good surface, the Positraction provides a quick and neat escape.

Back on the highway, potholes and road imperfections were soaked up by the variable-rate springs. The Sting Ray's low unsprung weight and suspension geometry had permitted GM to use relatively soft springs resulting in a comfortable ride with very little of the chop and pitch characteristic of a short-wheelbase car.

To the right of the steering column sat a 7000 rpm tach. For the 250-hp engine, a band of red on the tach face warned the driver that the needle should remain below 5000 rpm to avoid the embarrassment of searching for scattered engine parts. Between the speedometer and tach and just below them was the trip odometer. A small knob beneath the dash permitted the mileage to be reset when desired as long as it wasn't set while the car was moving. Above and between the two gauges were three more small windows: if the handbrake was left on, then the word "brake" would appear; the other reminded the driver that his headlights were on but not rolled upward. The third window was the high-beam indicator. The turn signal indicators were situated on either side of the panel.

To the left of the speedometer and next to the headlight switch were the fuel gauge and below that, the battery gauge, to the right, coolant temperature and below that, oil pressure. Below the dash were hood release, headlamp motor switch, pull cables for the kick panel-mounted vents, the knob for the coupe's exhaust vent and fan and to the right of the column, the handbrake. On the panel above the gearshift is the standard clock and below that, four control knobs: one for the heater, one for the defroster and the other two for the radio. An ashtray sits beside the gearshift lever and behind that, the shifting pattern is embossed in aluminum. When equipped with power windows, ($69) the switches are also located on the console.

I felt at home behind the wheel. As the sun dropped from sight, the lights were raised from their hiding place in the fenders. The movement of the beams across the road indicated that one light was just a bit slower than the other but a lubricant would improve its operation. As the road tests had reported, I noticed that light reflected from the instrument faces but not violently, the dished glass in the gauges was probably a contributing factor. At night, the 3-lane highway was a dangerous route. Sleepy weekenders drifted between lanes and the impatient drivers darted in and out of line. Avoiding the sleepy ones was advisable and fortunately at 70 mph the Sting Ray still had power to pass and quick, precise maneuverability for emergency reactions. For us, the car's quick steering could also be considered a disadvantage. It didn't take much wheel action to move that car and a drowzy moment could put us into another lane so the Sting Ray demanded attentive driving.

The trip was memorable and during that summer we made a number of other treks to the shore, one in particular coming to mind. My friend, Crazy George, worked on his dad's fishing boat and made plenty of money doing it—enough to pay cash for a brand-new '65 Sting Ray convertible. Milano Ma-

1. Styling stood fairly pat for '65 except for the three new vertical louvers in the front fenders, but some optional niceties lurked under the hood.

2. The run-of-the-mill 327 engines (in 300, 350, 365 and 375 hp sizes) had a new big brother for '65; the L-78 396-in., 425 hp Turbo-Jet—but you had to take Positraction and the transistorized ignition to get it.

3. A droopy-eyed '65 coupe gets the once over in, of all places, Dar es Salaam, Tanzania, as the United States sends abroad an exhibition featuring its plastic products.

4. Refined instrument backings helped solve complaints about glare and reflections.

THE BETWEEN YEARS: '64 TO '67

room with a black interior, the car was beautiful and, equipped with Chevy's new 396-cu.-in. big block, the car was one of a few announced at mid-year.

It was Friday night and we had just completed another city-to-shore drive. After eating, we were sitting in the coupe sipping root beer and visiting with the girls in the next car when the windows began vibrating. It had to be George. No one else on Long Beach Island had quite as distinctive a calling card. The deep rumble of 425 horses exhaling through side exhausts sidled up next to the coupe. Sure enough, behind the wheel of that incredible maroon machine was George, smiling affably.

Although a new factory option ($132), those side exhausts were designated by the factory for off-road service only in making the purchase, the driver accepted full responsibility for any loud-exhaust tickets that he might receive. Even now, the exhaust note was loud and considering its design, it wasn't surprising. The pipes bolted to the headers through a flange and running outboard of the body they clamped to the frame. The pipes were chromed only at the tips and ventilated, double-layer aluminum shields covered the rest. A fluted housing surrounded a chambered pipe and within, muffling material tried in vain to quiet the noise. I had seen George's car before but only in passing, so now I took the time for a good look. Externally, the biggest difference between the two was in the front fenders. In place of the phony engine vents in the '64 fenders, the '65 boasted real ones. The depressions in the '64 hood were remnants of the '63's fake grilles but for '65, the indentations were gone. With the exception of the traditional center bulge, the hood was smooth. The number of horizontal grille bars had been reduced to three which, instead of being chromed were now black; however, the grille surround remained chrome. The rocker moldings were simplified even further on the '65, reduced now to just one black stripe across an aluminum field. New wheel covers retained the 3-prong "knock-off" spinner but adopted a spoked look similar to that of the Cragar "mag" wheel. To make room for the new engine, GM designed the 396 hood with a "power dome." Using the basic outline of the small-block hood, designers flared the bulge out and up and on each side included a functional chrome vent for air intake and engine ventilation—it really looked mean.

Three new paint colors were announced for '65: Glen Green, Milano Maroon and Goldwood Yellow. Three earlier colors got different names and/or slightly different tints: Riverside Red became Rally Red, Silver Blue became Nassau Blue, Satin Silver became Silver Pearl. Two '64 colors were dropped; Daytona Blue and Saddle Tan. Tuxedo Black and Ermine White though, were continued. Inside, seats were again available either in vinyl or optional leather in black, red, blue, saddle, silver and white, and two new colors, green or maroon. Depending upon the color of interior, a buyer could specify either black, red, medium blue or gray carpeting and black and red or medium blue instrument panel. As in '64, convertible tops came in black, white and beige.

In addition to new colors, there were other interior changes for '65. Replacing the thin vertical pleats of the '64 seats was a plusher look of wide horizontal sections. Climbing into the passenger's seat, I noticed more padding around the edges of the seats, which provided greater lateral support than John's seats. The new door panels were one-piece molded vinyl rather than upholstered. The armrest for '65 was integral with the panel and not removable as before. Although it looked good when the armrest wears out, the owner must replace the entire panel. In John's coupe, the seat belts were always getting tangled and caught in the seats but looking at George's car, I was happy that the belts were kept in place with retractors and console-mounted clips. For '65, the same instrument layout continued but the gauges received a cleaner appearance. Conical for '64, the inside faces of the gauges were flat for '65. The faces remained black while the bucket outboard of the glass went from chrome to flat black. The '64 ignition switch had a

"lock" position permitting the driver to lock the ignition and remove the key but also included was an "off" position that allowed the driver to remove the key in its place and use the little protrusion on the switch to operate the ignition. Consequently, some owners developed some bad habits and instead of turning the switch to "lock," they would leave it in the "off" position, pocket the keys and walk off returning only to an empty parking spot.

The steering wheel appeared to be the same with a new chrome hub beneath the horn button; the new telescopic wheel option. By loosening the hub, the column moved in or out a total distance of 3 ins. Combined with the seat adjustment, it suited most drivers. However, the same adjustment was standard equipment on '63 and '64 models only it required a wrench and two people. At $42, the new option was obviously more convenient. Another optional first was a genuine teakwood steering wheel ($47) but considering the 15-week wait, George nixed the idea and settled for the walnut-grained plastic rim. For collectors, the teak wheel was to become a rare item. Federal safety rules considered it to be potentially dangerous, so it was pulled from the option list early in production. Another unusual option was the aluminum-case 2-speed Powerglide transmission ($195) and those who ordered it with their '65's got the benefit of straight-line shifting rather than the wiggly pattern of '64.

The biggest news for '65 was not outwardly apparent; 4-wheel disc brakes had become a reality. Offered as standard equipment, the disc brakes were well worth the wait. George loved them and so did Motor Trend (October '64): "During performance testing, the brakes were subjected to just about every brake test imaginable —and then some. One of the engineers' favorite tests was a series of 20 .70%-G (23 feet/second/second) stops from 100 mph. All stops were made consecutively, with no time between for cooldown, and each stop had to be made with absolutely no wheel skid. Existing drum brakes are lucky to survive one stop like this, but the Corvette discs were able to make all 20—time and again—right to spec. The discs' resistance to fade is pointed out by the fact that normal hydraulic-line pressure in this system (at 100-pounds pedal pressure) is 500 psi. After the 20-stop series, line pressure had risen to only 625 psi, a very nominal increase. It can honestly be said that this system is fade-proof. During other tests, the brakes have pulled test Corvettes down to swerve-free, no-skid stops at a deceleration rate of one

1. The Corvette for '66 saw two new engine options, a 390 hp 427 (L-36) and a 425 hp, also a 427. The 396-in. engine was discontinued. RPO N-14 was yours for $131.65; the side-mount exhausts. Listed as on off-road option only, this meant you took your own chances with your State's exhaust noise limits.

2. The Sting Ray's styling was holding up well in this, its 4th year, as sales climbed to 27,720 units, of which some 36 percent (not quite 10,000) were coupes.

3. An egg-crate grille replaced the earlier thin, horizontal bars.

THE BETWEEN YEARS: '64 TO '67

G—32.2 feet/second/second—a feat that has been believed physically impossible." Sports Car Graphic (October '64) reported: "After experience with the drum/sintered-lining setup—a previous HD option—we found this brake a distinct pleasure to operate, especially as the stopping potential is even greater. Repeated stops from above 100 mph produced no deterioration in braking efficiency and over 20-foot decels could be made with hands off the wheel."

Patterned after the Girling design used on Corvette's Grand Sport race cars, the new brakes were produced by GM's Delco-Moraine Division. Per wheel, the system consists of a 4-piston brake caliper mounted on a cast-iron rotor. Measuring 11.75 ins. in diameter and 1.25 ins. thick, the disc is riveted to the car's hub and rotates with the wheel. It is machined on its outboard and inboard sides to form contact surfaces for the flat brake pads in much the same way that conventional curved brake shoes contact the curved machined surface of the drum. Radial webs are cast between the two surfaces and create wedge-shaped air spaces. In addition to reducing the overall weight of the disc, the slots trap air to ventilate the contact surfaces.

Like conventional brake wheel cylinders, the 2-piece caliper contains two shallow cylinder bores in each half and within each cylinder; a spring, a rubber seal, a metal piston and to hold it all in, a rubber piston boot. Sandwiched between the two caliper halves are two large asbestos-lined brake pads. The pads attach to the caliper halves with a retaining pin. When bolted together, the caliper straddles the disc so that the lining of one pad contacts the inboard side of the disc and the outer pad contacts the outboard side. Lines run from the split-chambered master cylinder to each of the calipers. Unique among contemporary designs, the Corvette system is designed for zero clearance between the pads and disc. Instead of retreating from the disc upon brake pedal release, the Delco pads remain in constant contact. It's only a slight contact, however; enough to keep the surfaces dry and clean. This permits instant brake response under all conditions and failed to produce rapid wear as predicted. GM had carefully chosen the proper pad material and under testing, it survived a minimum of 50,000 miles. Large pistons in the front and smaller in the rear produced a 65/35% distribution of braking front to rear. Consequently, the rear pads last much longer; in fact, the projected wear rate for the fronts is 57,000 miles and 127,000 miles for the rear with no disc wear. Much of this longevity is due to the large pad size and compared to the standard drum brakes, the pads provide 461 sq. ins. of swept area (the area contacted by the pad or shoe) and the drums, only 328 sq. ins.

For those who wanted them, conventional drum brakes, front and rear, were available as a $63.15 credit option. Reportedly, only 316 drum brake '65's were built and those are prized today. Apparently, the heavy-duty metallic brake system continued to be offered at its outrageous price of $629 but no reports are available on numbers sold.

HORSES FOR COURSES

Now I was riding with George, staring at the speedo and watching it climb rapidly to 100 mph, 110, 115. I tried to be cool as I held on tightly, but the continued pull pressed me to the seat; 120, 125. I had blown my cool but it no longer mattered; 130, 135, "George, slooow down."

At 140 mph, when the rails on a bridge were merely a white blur and a traffic light loomed a ½-mile ahead, George backed off and pumped the brakes. John's coupe was blocks behind but coming up fast. A couple of miles back, John and George had agreed to a contest. For a 250-hp Vette, John's car was amazingly quick but no comparison to the 425-hp honker. To make things a little more even, it was decided that George would start off in third, and, starting in first, John would get a three-car lead. It didn't take long for that big block to catch the coupe. No contest. Despite George's good-driver reputation, I was glad it was over. To have gone that fast in one of the first 396 Vettes was a memorable experience.

Detroit was in the middle of a horsepower race and to GM, the time was right for a big-engined Corvette. Powered by GM's mysterious 427-cu.-in. V-8, a Chevy stock car broke NASCAR records during its first appearance at Daytona in early '63. Before it could race again, corporate powers said "no more" and the engine remained a mystery until 1965 when the Mark II 427 became the Mark IV 396. In January of 1965, production began on the Turbo-Fire 396. With a bore of 4.09 ins. and a stroke of 3.76 ins., the 396 featured five large 4-bolt mains; 11:1 compression, solid lifters, 2.19-in. intake valves, 1.72-in. exhaust valves, a large Holley 4-bbl. carb, a fully transistorized breakerless ignition (RPO K66), strong impact extruded aluminum pistons with heavy connecting rods, cross drilled main bearing journals for 360° oiling to the rods and a unique set of cylinder heads. Like

the 427 NASCAR engine, the 396 was dubbed the "porcupine" because of the valve stem angles. The intake valves angle toward the intake port for a more nearly straight flow of air/fuel mixture. Although at a more conservative angle, the exhaust valves tip away from the intakes so they can be closer to the large port for better mixture flow. The combustion chambers are a modified wedge design with the spark plug placed in the center for a more uniform burn. The result is an engine that can produce in stock form 425 hp at 6400 rpm and 415 lb.-ft. of torque at 4000 rpm.

With accessories installed, the 396 V-8 weighs approximately 680 lbs.—about 150 lbs. more than the 327. To accommodate the 396, the Corvette was equipped with a wider radiator, a bigger fan, a small, high-output water pump and a stronger clutch plate sometimes criticized for its excessive pedal effort. To handle the extra engine weight, GM installed stiffer front springs, a ⅞-in. anti-sway bar for the front and a 9/16-in. bar for the rear. The rear axle shafts were made of stronger steel and the U-joints shot-peened as a deterrent to metal fatigue.

Equipped with the 327, the '65's front-to-rear weight distribution was 47/53% and with the 396, 51/49. However, test reports indicated very little difference in handling between the two. Available only with the close-ratio 4-speed and Positraction rear axle, the 396 Vette was comparable to the fuel-injected model in 0-60 times. With the standard 3.70:1 rear axle, the big block required 5.6 secs. to 60 mph and the fuelie achieved it in 6.3 secs. Top end was better for the big block with a ¼-mile speed of 103 mph in 14.1 secs. while the injected engine ran the quarter in 14.6 secs. with a terminal speed of 98 mph. Fuel consumption for the larger engine was in the 9-12 mpg range, for the fuelie, 14 mpg.

For brute power, George's '65 was impressive but around town, it was a temperamental beast requiring frequent stops at the gas pumps. For '65, there were five engines from which to choose: Replacing the 365-hp/327 in mid-year was the 396-cu.-in./425-hp 327/250-hp, 327/300-hp and continuing through the model year, the 327/375-hp fuelie. The fifth engine was a new entry to the '65 lineup. Producing 350 hp at 5800 rpm and 360 lb.-ft. of torque at a fairly low 3600 rpm, the new 327 was essentially the old 365-hp unit with a milder cam and hydraulic lifters. Performance is similar to the older engine but without the noise, valve adjustments and fouled plugs. Through cam timing, engineers were able to "streetify" the engine to its present state; that is, with peak power and torque in the more usable rpm ranges. With its aluminum finned valve covers, chrome air cleaner, 5-qt. pan, 11:1 compression and Holley 4-bbl., the 350 was a $108 option.

The base '65 convertible with 250-hp engine, 3-speed synchro trans and 3.36:1 rear axle, retailed for $4022; the coupe, $4233. Optional equipment included Four-Season air conditioning ($413) for the 250-, 300- or 350-hp engines; 3.08 economy axle (without Posi) when ordered with the 4-speed and two smallest engines ($2); Positractions in all ratios (3.08, 3.36, 3.55, 3.70, 4.11 or 4.56) for $42, 300-hp engine ($53), 350-hp engine ($105); 365-hp V-8 ($126); 375 hp with fuel injection ($527) and 396 with mandatory Posi and transistorized ignition ($286).

NEW LIFE—NEW STING RAY

The following summer, John and I went separate ways—the Navy for him and the Army for me. I spent a year of service in Vietnam and each month I'd study the car magazines, looking at photos of Vettes and remembering old times. I saved my money for the day that I'd finally go home and buy my own Sting Ray. Somehow, that day did arrive and within three months, I bought a '65 convertible, Milano Maroon with black top and black interior. The car was in fine shape and its 365-hp engine had recently been rebuilt and blueprinted. Between plug changes, it ran strong but the previous owner had installed chrome rings and they took forever to seat. Aside from fouled plugs, the only problem I ever had with the car was a stripped tach drive—a problem many Corvette owners have experienced. Naturally, I joined the local Corvette club and attended a number of sports car rallies. There was a fellow in the club with a '66. Comparing the two, I could see only minor differences. The '65 was a hard act to follow but then why mess with success? It had pleased most everyone with its new brakes, two new engines and cleaner styling and the numbers proved it—a record 23,562 '65 Corvettes were sold—an increase of 3562 over 2.6%. At a club meeting, we parked side by side and began calling off the differences: cosmetically, the most

1. At full cry on the straights at Milford, an early '66 gets a tester's workout. Refinement in the suspension areas, and potent optional powerplants, continued to hold the Corvette head and shoulders above any other American car in terms of performance.

2. Performance-minded buyers, especially those who wanted the ultimate, shopped their options carefully in 1967. The base coupe had an all-up weight of 3155 lbs (20 lbs. less than the convertible). But power windows added 5 lbs., air conditioning 93 lbs., power brakes 12 lbs., power steering 21 lbs., and 21 lbs. more for the AM/FM radio. The 427-in. engine meant another 175 lbs., but deduct 73 lbs. from this for the optional aluminum heads. Leave out the heater and save 20 lbs., and remember that the 4-speed trans was 5 lbs. lighter than the 3-speed.

3. The Corvette took easily to any road and, it was always fun to drive.

THE BETWEEN YEARS: '64 TO '67

identifying change was a new egg-crate design grille that replaced the 3-bar grille of '65. The '66 rocker molding was finely ribbed like the '63; the coupe gave up its exhaust vents in the roof pillars and new wheel covers resembled the '65 covers. Inside, changes were greater. The '65 seats weren't holding up—they seemed to be ripping at the center seams in the cushion. For '66, the vinyl was expanded in narrow horizontal pleats. Inside, door handles were chrome and dash controls were chrome with flat black centers. As in '65, the big news was mechanical. Engine selection was reduced to only four: the 300-hp/327, the 350-hp/327 and new for '66, the Mark IV 427 in two forms. The 250-hp engine was replaced by the 300-hp engine as the base powerplant. Sadly, the fuel-injected engine did not return in '66. At $527, it was about $200 higher than the biggest 427 and for that reason, GM felt that it was too expensive to keep. The 396 was dropped after one year and with a larger bore (4.25 ins. vs. 4.09 ins.), returned as the Turbo-Jet 427. The high-performance version of the 427 was rated conservatively at 425 hp at 5600 rpm and 460 lb.-ft. of torque at 4000 rpm, but more honestly, it was closer to 450 hp at 460 lb.-ft. of torque at 3600 rpm. This reduction in power was achieved by reducing the size of the intake valves, cutting its compression from 11:1 to 10.25, using a hydraulic instead of a mechanical lifter cam, a smaller venturi carburetor on an iron rather than an aluminum manifold, cast instead of impact-extruded pistons and two rather than four-bolt mains.

A new 3-speed from Saginaw became the base transmission. Synchromesh in all three gears, the new box has ratios of: 2.54:1 first, 1.50:1 second and 1:1 third. The wide-ratio and close-ratio 4-speeds continue into '66. At $4295 for the coupe and $4084 for the convertible, the '66 Sting Rays are only slightly up from '65. With the exception of the engine changes and six new options, the option list is the same as '65 and surprisingly prices are too. The new options include an air-injection reactor (smog pump) for California cars ($45), shoulder harnesses for driver and passenger ($26), headrests ($42 per pair), a 4-way flasher switch ($12), a heavy-duty brake system (J56) for the 425-hp engine with power brakes which includes: stronger brake hoses, larger sintered brake pads with nickel alloy backing plates held in place by two retaining pins instead of one and an iron brace for each of the front calipers. The price? $342. Updating the F40 suspension option of '65 is the F41 package. Sold only with the 425-hp engine, the system includes stiffer non-variable-rate springs front and rear, bigger 1 ⅜-in. diameter shocks, a heavier 15/16-in. front anti-sway bar and a bar in the rear. The suspension option goes for a mere $37.

For '66, two new paint colors are added to the '65 list and two are replacements: Laguna Blue and Trophy Blue and, replacing the '65 Glen Green, is Mosport Green. Replacing Goldwood Yellow is the new Sunfire Yellow. In interior colors, green and maroon were dropped after '65 and bright blue was added.

SUMMER OF MY '67

I kept that '65 for almost a year but then one summer evening, I was forced off the road, across a lawn, through a rock garden, into the woods and a telephone pole. Totaled! My beautiful car totaled. Fortunately, the humans survived injury but the Corvette was parted out to the highest bidders. The insurance company paid top value, and with $3600, I bought a very clean, low-mileage '67. It was a Rally Red convertible with black top, black interior and a 427 black hood scoop. Beneath that hood lived a 350-hp 327 but who had to know? I have enjoyed the crisp performance of that car with its 3.70:1 rear axle, dependable, low-maintenance engine and I've particularly enjoyed the curious looks at that hood. The economy of the 327 with the prestige of the 427.

Changes for '67 are subtle; in fact, refinements would probably be a better word. The new '68 body style was supposed to appear for the '67 model year but delays forced the creation of the '67—perhaps the nicest Corvette ever produced. Five smaller vertical engine vents replaced the three used in '65 and '66. When combined with the new flat black rocker panel moldings, the shorter, more delicate vents seemed to make the

car lower, sleeker, and not nearly as bulky as the '66. From the rear, a large backup lamp above the license plate identifies the car as a '67; new 6-in. silver-painted wheels with five slots each looked great with the small hubcap and chrome beauty rims—they no longer look fake. A column-mounted 4-way flasher becomes standard equipment for the '67 and signalling a lane change no longer requires a definite turn signal. The handbrake lever was redesigned and moved between the seats. Instead of the T-handle, a single black plastic lever with finger grooves and a release button on top does the same job, but more conveniently. A new hood dome redesigned to clear the 427 air cleaner is closed to ventilation but with a couple of tools, the front inlet panel can be removed for improved cooling. The dome and the traditional center bulge are painted black and pinstriped. In place of the lift-release seatbelt buckles, the new belts have a pushbutton release. The front seats now have latches to lock the seatbacks in place for collision safety. The old 3-prong knock-off aluminum wheels have unfortunately lost the spinner (a safety regulation) and instead, a flat hub cover conceals conventional lug-nuts. The price for the turbine-design wheels has been reduced; from $316, in '66 to $263 for a set of five. Another important safety item—energy-absorbing steering column—is a standard item for '67. The horn button, emblems and fuel cap cover have all been redesigned. A larger rearview mirror with padded frame is standard and now available for the removable hardtop, an optional vinyl roof cover. At $4353 for the coupe and $4141 for the convertible, the '67 Sting Ray is just $58 higher in price than the '66. Except for the engines, the options list is basically unchanged. However, in paint colors, Glen Green has returned to replace the silver green of '66 and with it returns the dark green interior.

Engine selection for '67 consists of three returnees: the base 300-hp/327, optional 350-hp/327, 390-hp/427 and two new big blocks: the 400-hp/427 and a 435-hp/427 each with three 2-bbl. carburetors. Producing 400 hp at 5400 rpm and 460 lb.-ft. of torque at 3600 rpm, the smaller of the two has 10.25:1 compression and hydraulic lifters. The bigger engine has an 11:1 compression, high-lift cam, solid lifters and transistorized breakerless ignition. This engine produces 435 hp at 5800 rpm and 460 lb.-ft. of torque at 4000 rpm. Under a triangular air cleaner are three of Holley's 2-bbl. carburetors with a mechanical/vacuum controlled linkage setup.

The epitome of all Corvette engines appeared in the spring of '67; the L-88. Featuring aluminum cylinder heads and a huge list of performance equipment, the L-88 427 engine was once rated at 560 hp. Sold sans emissions equipment, the L-88 was authorized for street use only in those states without strict smog laws. With a compression ratio of 12.5:1, there are very few stations dispensing the proper 103 octane fuel. Apparently, only 20 aluminum-head L-88's were ever produced and only three are known today. Equipped with a large hood scoop, an 850-cfm Holley 4-bbl., on an aluminum manifold, a super high-lift cam, special valve springs, caps and rocker arm balls, heavy-duty clutch and smaller flywheel and heavy-duty aluminum radiator and many more internal modifications guarantee ¼-mile times in the 13's and 0-60 times around 5 secs.

The '67 Sting Ray has often been called an interim model between '66 and '68. It was a year of many small refinements to body and interior and as the last year of that body style, I personally believe it's the best Sting Ray ever produced. Of course, I am a little partial but until a finer car comes along for the same money, I think I'll just keep mine forever, and ever and ever . . .

1. Corvette year-spotters were helped in '67 by the obvious addition of the centrally-mounted back-up light, the only year that featured it.

2. The aluminum heads not only trimmed poundage off the 427, but aided breathing through revised combustion chambers, and hard valve seat inserts and guides allowed high revving.

3. The 427 topped with three 2-bbls. is identified by the triangular air cleaner, and it's as much of a joy to behold as it is to climb in and stand on it!

4. A high-dollar restoration, this '67 is much-desired today as the final year of the classical Sting Ray body. While the name would be carried forward, (but changed to the single appellation Stingray for 1969) purists today think only of the '63 - '67 Corvettes as "real" Sting Rays.

1963 Corvette Sting Ray
Split Window Coupe

The first Sting Ray was a show car in production, a racing car on the street . . . and a promise fulfilled

BY ALLAN GIRDLER
PHOTOS BY RICHARD M. BARON

WHEN THE FIRST Chevrolet Corvette Sting Rays appeared in public, we public held our collective breath: This was all too good to be true. Like the proverbial kids, we'd spent a lot of time with our faces pressed to the candy store window. We knew there had been racing Corvettes since the model appeared in 1953, but we also knew parent General Motors didn't like racing. We knew some really new and different cars had been shown by Chevrolet, but show cars seldom made it beyond center stage.

And deep down in our secret hearts, we knew that the Corvette, America's Only Sports Car and all that, was fast but not terribly sophisticated; was, in fact, a cut-down passenger car with hot-rod engine and suspension and wasn't quite what would be known much later as State of the Art.

Then, in the fall of 1962, all the promises Chevrolet had made to the sports car crowd came true, all at once, which is why it was difficult to believe it then and why it matters so much even now.

But we'll begin with way back when, in the late Fifties. GM had allowed some farsighted men to build the Corvette, which at first was more personal car than sports car but later got a healthy V-8 engine and the ride comfort of a truck. They stayed with the fiberglass body, something that was at first supposed to be an experiment, and they stayed with two seats when rival Ford had gone the quick way to profit with four seats, and they'd allowed some suspension and brake and engine work, notably fuel injection and a 4-speed manual gearbox, that kept the Corvette on the track if not always first across the finish line. But the Corvette as sold to the public was big for two people and it was, not to mince words, crude. The live rear axle hopped and tramped under power and the body was a sort of old shape with gobs of new bits, e.g., four headlights, stuck on it.

Meanwhile one Bill Mitchell, then head of GM's styling staff, somehow managed to get the title to a racing Corvette chassis, a special one and not much like the production version, signed over to him. Heck, yes, it was a political thing, and it probably didn't fool anybody but the point was Mitchell was allowed to have the best guys in the shop work on his racing car, all for experimental purposes and/or as a hobby, ho ho. Mitchell did the body styling himself and because he'd been inspired by deadly denizens of the deep or simply liked the sound of the name, his special Corvette was called the Sting Ray.

Meanwhile, Number Two, Zora Arkus-Duntov, head engineer for Corvette, had been given the dream assignment he somehow managed to get every few years, that is, he was told to design a new and improved Corvette.

The resulting project, code-named XP-720, was much more new than it was improved. Virtually the only components carried over from 1962 to 1963 were the engine and transmission, which was fine as they were the best parts of the old version.

But XP-720 began with a new frame, stronger and more resistant to twist, yet lighter. The front suspension was updated; okay, it was passenger-car parts but they were new ones where the previous Corvette had used suspension components from much older cars. There was new steering, quicker and lighter at the same time, and, as part of the design, the control arms could be switched and the ratios changed for really quick steering.

But the major difference was the rear suspension. Independent. Yes. Honestly, fully independent suspension just like the upmarket European cars used, and the sort of thing the buyers of said imports scoffed that dumb ol' Detroit would never build . . . except here it was, at last. The 3-link system used the halfshafts as a location device, and there was a single, transverse spring. Duntov thought the arrangement perhaps old-fashioned, and indeed such a spring had been seen on the Model T Ford, never mind the Cobra. But it all worked.

The Corvette engine was the 327 V-8, grown from the 265 and 282 versions that gave Chevrolet its performance image in the Fifties. Corvette buyers had a choice of engine version, as the 327 came with a mild camshaft and hydraulic valve lifters and a rating of 250 bhp, or an optional higher compression but mild cam and 300 bhp, or the solid lifters and radical cam timing with alloy intake manifold and giant 4-barrel carburetor, 340 bhp, or the topline, take-no-prisoners 360/327 with the hot cam and fuel injection. This wasn't a completely free choice, as the buyer could have the 4-speed stick with any engine, but the 2-speed automatic came only with the two mild versions. Oh, and back before emissions certification and fuel economy standards (freedom is what we called it then), the buyer had a choice of internal ratios for the stick shift, and of final-drive ratios.

This mechanical freedom and experimentation probably wouldn't have mattered much except for the way the new car looked.

First, it looked different. Mitchell's, um, private racing car had been a good racing car, plus it was sleek and slick and had its own shape. By no mistake, the 1961 and 1962 Corvettes, which were offered only as convertibles, had the rear body sections that first appeared on that racing Sting Ray. The backs were much cleaner than the fronts, but what mattered was that they established how a modern racing car was supposed to look.

With the 1963—the full Sting Ray, one could say—the look was complete. It was different, with a high horizonal knife edge and an air intake almost concealed below what looked like an aerofoil.

The racing Sting Ray naturally didn't have headlights, so it was easy to have that high, frontal edge cutting the air. For the production version, though, keeping that look meant concealing the headlights. They were mounted in rotating buckets and twirled in or out of sight by electric power.

Next, the new car came as a traditional convertible, with or without a removable hard top, or as a coupe. This last was new for Corvette although other sports cars, as in Jaguar and MG, had offered fully enclosed models earlier.

The difference was a matter of emphasis. The other sports coupes were closed versions of open cars. (We exempt the Jaguar E-Type here, because while it appeared before the Sting Ray did, the Corvette was in final form before the Jaguar went public.)

The Corvette coupe was its own self, so quickly recognized that the design seemed much simpler than it was. But Mitchell had taken the cutting edge, the high sharp look of the racing roadster, and blended it with a teardrop, the pointed tail of nature's slickest shape. While the race car's shape made it a handful at top speed, the production version was revised in a wind tunnel—shades of progress to come!—so there was downforce on the front at speed.

For character, so to speak, two odd parts: First, to keep the fiberglass body strong and light at the same time, the body itself had only two holes, the doors. The rear window was solidly mounted and there was no trunk lid or hatch or rear access of any kind. The spare tire had its own bin, below the floor, and all stowage, as in kids, picnic supplies, luggage, groceries, etc, went in and out through the doors and over or behind the tipping seats. The coupes had sort of a cave; in effect, a flat floor with sloping and narrowing roof. The convertible had a folding panel, behind the seats and hinged at the rear, that raised up to admit the folded top. When the top was up, there was room for a suitcase or two under the lid. When the top was down, there was room only for the top.

The second odd part was that Mitchell stressed the unity, the solidity of the teardrop rear section by having two separate rear windows, small ones, one on each side of a solid center panel running down the middle of the roof. The coupe became known as the Split Window coupe about as soon as the 1964 models were introduced with the one, larger rear window. Mitchell liked the design because it worked when you looked *at* the Corvette, while Duntov objected because it didn't work when you looked *out of* the car. Duntov was right in the short run, but Mitchell's trick made the 1963 Split Window coupe one of the most easily recognized models of recent times. But that, of course, is now.

Back then the new Corvette was simply sensational, the sort of new car enthusiasts actually went to showrooms to see, then exclaimed over when the cars appeared on the road.

The new car looked terrific. Equally, it was fast. Just how fast naturally depended on the engine and gearbox in the test example; that is, there was a lot of choice and a big jump from 250 bhp with automatic to 360 bhp with 4-speed stick and drag-strip final drive.

Road & Track tested a 1963 convertible with the 360 engine, the 4-speed and the 3.70:1 final drive fitted as production with that engine and transmission. R&T's companion publication, *Car Life*, tested a coupe with the 300 engine, automatic transmission and 3.36:1 differential.

Remarkably, the two cars were nearly identical in weight, 3030 lb at the curb for the convertible, 3048 lb for the coupe. So the times: 0–60 in 5.9 seconds with the hot engine versus 7.2 seconds for the mild one, quarter miles in 14.9 or 15.5 seconds, with trap speeds of 95 or 86 mph. Neither magazine actually ran its car for top speed, although they estimated 130 mph for the calm coupe, 142 for the racy convertible. Surely the more streamlined coupe would have a higher potential top speed, assuming the hot engine and the proper gear, but even back then, before the imposition of a national speed limit, such figures were more academic than useful.

The real benefits from the new Corvette, though, were in—gasp!—comfort.

Different kinds of comfort. The new Corvette was actually smaller than the old one on the outside, but larger on the inside, simply because the engineers had been given more freedom to make their own components instead of using what

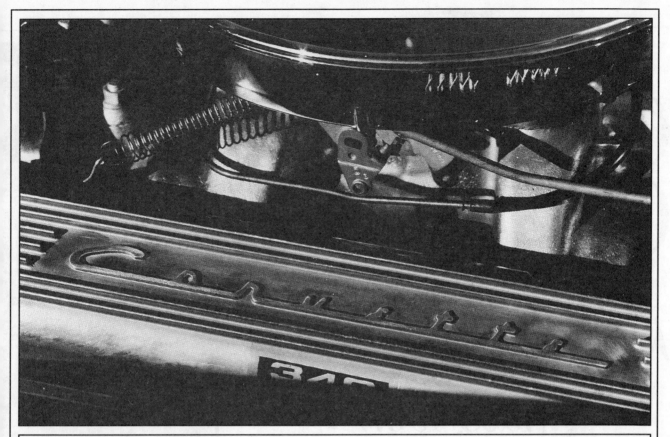

the parent division had in stock. And these were the guys who'd actually buy such a car, so they were careful to improve the head room and give more leg space and provide an adjustable steering column, so there was room to relax. And because the steering was lighter as well as quicker, the driver didn't need to heave on the wheel with his shoulders and thus could be farther away from it.

Next, racing designers were just then working out that if you provided the correct geometry for the wheels when they went up and down, and the correct control, you could let them move more. Translation: softer springs and better shocks. Plus, there was that independent rear suspension. It wasn't a gimmick. The Sting Ray Corvette did handle better, on smooth roads as well as rough ones. If this came from good design and proper control of location points and roll centers rather than the principle of independent suspension, well, heck, it gave the owners something to talk about.

It's also worth noting out of time frame that back then, when the family sedan was tall and square, being low and sporting didn't mean so low as to impose on the occupants. So Corvette drivers and passengers could get in and out with a minimum of contortion.

Just as the Split Window Coupe was an instant classic, so was the Sting Ray Corvette an instant success. Sales were up 30 percent for that year and would have gone even higher except that the plant couldn't supply all the cars ordered. There was a waiting list—don't you reckon Chevrolet would love to hear those words again?—and dealers weren't inclined to deal. Early predictions were for the coupe to attract most of the buyers, while the convertible was expected to go to the races. As it happened, probably because it was easiest to schedule production 50-50, production totals for 1963 were 10,594 for the coupe, 10,919 for the convertible.

The example shown here is too good not to be true, and the owner is a man who should be disliked for doing what the rest of us wish we'd been smart enough to do.

When Rick Costa was a kid, he saw the first Corvette and vowed some day he'd have one. So when he was in college and about to become a father, Chevrolet introduced the Sting Ray Corvette. Costa bought one, the top of the line with fuel injection and all the good parts.

Costa raced it in autocrosses and solo events and won. The Costa babies came home from the hospital in the coupe. When *Corvette News* named the Top Twenty Corvettes in the U.S., this one, between trips to the store and runs on the track, was on the list.

Okay, be fair. Costa has had other cars, a 1965 Corvette for when he got serious about racing but didn't want to carve up the 1963's body for wide tires, and since then a 1985 Corvette for daily driving.

Point here is this car has done 119,000 miles. It has not been restored or rebuilt. Instead, Costa has maintained the car to better-than-book standards. The oil is changed often, the car isn't allowed to get dirty but when it does, the cleaning process is also an inspection tour, and so forth. Perhaps the real secret is that Costa does all the work himself, so he knows it's been done and done right.

Nor does the owner find this odd. Rather, he's a bit tired of hearing from all the guys who wander up and say, "Gosh, a Split Window Coupe! I sure do wish I'd kept mine!"

No doubt they mean every word. But if ever there was a timeless design, if any Corvette is *the* perfect example of what the car is supposed to be, do and represent, the Sting Ray Corvette, Split Window Coupe version, has to be that car.

The surprise shouldn't be that this is a one-owner car.

The surprise should be that any owner would have sold.

1963 Split Window Chevrolet Corvette, provided by Rick Costa, Laguna Niguel, California.

CLASSIC PROFILE

CHEVROLET CORVETTE *Sting Ray* (1963-1967)

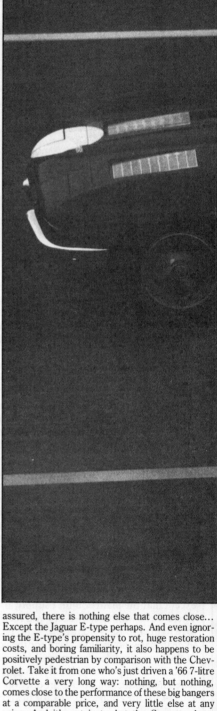

FOR *Performance, looks, reliability, longevity, parts availability, ease of maintenance and restoration, investment potential.*
AGAINST *Fuel consumption, build quality, detail design.*

There's a lot of old nonsense talked about Chevrolet Corvette Sting Rays in this country – a lot of waffle about how unrefined and bad mannered they are; about how they rattle and shake and stink of hot glass-fibre; even about how they aren't sports cars at all. Well let's get things straight right away: the Chevrolet Corvette Sting Ray, new for 1963 and continuously developed through to '67, is one of the best value for money sports cars on the market these days, and is unjustifiably maligned by the ignorant European enthusiast.

Sure Corvettes lack sophistication; sure they lack the attention to detail so important to classic car investors and enthusiasts, but let's not be blinded to its qualities. And it's only we Europeans that are. Sting Ray prices have virtually doubled in America during the past nine months. The Yanks have laughed all the way to the bank as they've shipped all their old Jags and Healeys back to Blighty while recognising the value of their own fine products. Sting Ray fever is rife in the US. You wanna drag? You wanna hurry...

Just consider the alternatives for a moment. What other sports car readily available for less than £20K is as good-looking, long-lasting, reliable, substantially built, cheap to maintain and restore, easy to restore and run, exceptional to drive, and as thoroughly cosmopolitan as the Sting Ray? Rest assured, there is nothing else that comes close... Except the Jaguar E-type perhaps. And even ignoring the E-type's propensity to rot, huge restoration costs, and boring familiarity, it also happens to be positively pedestrian by comparison with the Chevrolet. Take it from one who's just driven a '66 7-litre Corvette a very long way: nothing, but nothing, comes close to the performance of these big bangers at a comparable price, and very little else at any price. And it's not just what the Corvette does, either. It's the way that it does it.

Then there's the Sting Ray's other great quality over its pathetic competitors: its multiple identity. Corvettes came in so many different specifications

that the cars' characters vary dramatically. There are Sting Rays for most occasions:

Between 1963 and 1967 (inclusive) 14 different engine specifications (varying in output from 250hp from 327cu in to well over 400bhp from 427cu in) were offered together with five different transmissions in both coupe and roadster Corvettes. Add to those variables the possibility of power steering and air conditioning fitment, to say nothing of a host of other options, and it becomes clear that the Corvette Sting Ray could be anything from poseur's boulevardier to out-and-out race car. These are also good reasons for Sting Rays being very variable in price today, because you can pay anything from £10,000 to £100,000 for a production Sting Ray. It just depends on your priorities.

Choosing your Sting Ray is like buying any other old fish – it's luck of the draw. The chances of your catching exactly the specification car you want is pretty slim. Compromise is inevitable, unless money is no object, but basic criteria must be considered. You must decide whether the versatility of the more expensive roadster outweighs the appeal of the elegant coupe. You must decide whether you require an automatic or a manual car. Then you must decide whether you desire the smaller capacity 'Small Block' version of your chosen model, or the incredible 'Big Block' type. And last but by no means least, you must decide whether originality is important to you, because Sting Rays with all-original mechanics are worth a whole lot more than cars with non-original engines. Which is great news for the enthusiast who couldn't care less whether his motor was kosher, but a bore to the investment purchaser who must buy what Chevrolet would have termed 'The Real McCoy'.

Technical Description

The Corvette Sting Ray was the result of continuous development of the Corvette from its initial introduction in 1953. At that time the 'Vette was nothing more than a dressed-up Chevrolet Saloon – and Chevy saloons were very boring in 1953. The introduction of the car, if commercially pathetic, did wonders for Chevrolet's utilitarian image, though. It was the first GM dream car, (following in the wake of such brilliant ancestors as the Y-job (1937), Le Sabre (1951) and XP300 (1952)), to have real production possibilities. Possibilities which were so practical that the 'Vette was being built in limited numbers just 15 months after its 1952 Motorama debut at the New York Waldorf Hotel.

Then came the Corvette performance race. The cars acquired V8 engines and manual transmissions in place of their old six-cylinder, automatic types, and power progressively increased. Simultaneously came suspension developments and restyles, and even Corvette-derived works racers – in particular the SS and the Sting Ray. It was the latter, a development of the aborted Le Mans SS, which first gave a hint of what the next generation Corvettes would look like. It was Bill Mitchell's creation – the successor of Harley Earl as GM's vice president in charge of styling.

The Sting Ray racer not only looked like the Sting Ray production cars to come, but it used many similar components, and was made of glass-fibre too. But there the similarities end. The Sting Ray racer had a tubular chassis, and although the production car was always envisaged that way, in the end it ended up with a steel box-sectioned affair. The new chassis was the work of the brilliant Belgian-born Zora Arkus-Duntov together with H.F. Barr and E.J. Premo. They drew from their experience in building the Corvette SS, original Sting Ray and mid-engined CERV-1 monoposto race car to create the new suspension/chassis mix.

The Sting Ray production cars used a conservative boxed/ladder-frame arrangement which swept its outriggers up over the rear axle. A separate steel reinforced frame then held the front door posts and framed the door apertures, even forming a roll-bar in the coupe, which was a good deal stiffer car than the roadster because of it. The new car's wheelbase was reduced by 6in over its predecessor and both the front and rear wheel tracks were reduced. Note too that the chassis ran directly below the doors, which kept the Sting Ray's seating low and offered thankful side impact protection too.

Suspension was derived substantially from CERV-1, but there was no room for coil springs to the rear of the Sting Ray so a single transverse leaf was used, mounted below and behind the rear axle. The upper wishbones were actually the rear driveshafts (which therefore have two universal joints) and stamped-steel radius arms kept the rear suspension in position. Front suspension was by double wishbones and coil springs.

The Sting Ray's centre of gravity was 1.5in lower than the 1962 car and with a full tank of fuel weight distribution is almost exactly 50/50 front to rear (the Big Block cars came later, remember). The Sting Ray was a little lighter than its predecessor too, although at 3012lb it was no featherweight.

Braking was still by hefty drums when the Sting Ray was launched (which were scarcely adequate to stop the 150mph car) and steering, in either 'quick' or 'slow' ratios, was a Saginaw recirculating ball arrangement which worked well. Power steering was optional, initially on the 250bhp and 300bhp cars. Tyres were relatively minuscule 15in diameter, 5.5in wide affairs on either steel or alloy wheels.

While it is easy to detail here the basic specifications of the original 1963 Sting Rays, the enormous number of specification permutations are best confined to tabular description (see below). Styling details changed every year too, so space does not allow for their description either. Remember these basics though: The faithful Chevrolet V8, ohv engine was offered initially as a 327 only, with manual or automatic transmission, and with a fuel-injection option too. The fuel-injection was last seen in 1965,
the year the 396cu in Big Block engine was introduced. The 396 only lasted for one year before the 427cu in Mark IV engine was launched which remained with the car as an option (in various states of tune) until the Sting Ray's demise at the end of 1967 (the restyled 1968 car was called the Stingray – all one word!).

If you want your Sting Ray to stop properly then a post-65 car it will have to be (when the disc brakes were introduced on all four wheels). And, lastly, remember that the classic 'split-window' coupe was only built for that first year – 1963. After that it was deleted, because its lovely central rear window vane conveniently obscured a police motorcyclist, among other things!

Driving Impressions

If ever the expression 'once driven, forever smitten' was appropriate to an automobile, the Sting Ray is it. Because until you've driven one of these extraordinary machines there is very little way you can appreciate just how exhilarating they are. Any preconceptions about 'Vettes, good or bad, are bound to be wrong. Only one thing can be guaranteed – Small Block or Big Block, the Sting Ray will impress at the least, more likely shock and, in Big Block guise, even reduce its passengers to laughing tears.

This is all believeable because although the Chevrolet Corvette Sting Ray was America's only true sports car, it came from a nation who work hard and play hard, who indulge in more sport than any other. So the Yanks not surprisingly got the vital ingredients right – ultimate exhilaration at the price.

No, Sting Rays are certainly not beyond criticism. Far from it. And that is precisely why they are still affordable. Just getting into one is a little disappointing. The standards of finish are very poor: lots of ill-fitting plastic, much of which won't have worn very well; poorly designed, cheap and nasty seats; tacky instrumentation, at least in the post-'65 cars; poorly-seated side windows; even an umbrella handbrake (on all but the '67s, but that model's central brake is very nasty too). And such low standards of design-quality, if not build-quality, are not forgotten on the open road where a Sting Ray roadster's scuttle shake can be atrocious, to say nothing of its leaking roof.

Such lack of attention to constructional detail shouldn't put you off though, even if a Jaguar E-type is a paragon of achievement by comparison. Much of the quality of the car will depend on how well it has been looked after or restored, and Sting Rays vary massively in this respect. Go for a coupe if panel fit and chassis-flexing bother you, and leave an average, bodged Sting Ray alone if you're a stickler for quality.

But perhaps we are missing the point. Sting Ray roadsters are all about kids having fun – summertime fun, and long care-free weekends away. And Sting Ray coupes are for taking motor racing, or for getting to the office on time. Coupe or roadster, Sting Rays are all about what happens to your body when you put your foot in the carburettor…

You sit low in a Sting Ray, with the dished steering wheel close to your chest. Pedals are big and well-spaced, the short chromium-plated shifter is an easy reach. Instruments are all visible, and the dual-cowl dashboard design is pleasing in its Batmobile connotations. The bonnet view is suitably spectacular, especially with the Big Block's hood to look at.

Having gassed-up the carburettor with a few pumps of the organ-pedal throttle the 327 fires up straight away and runs quietly. Warm it up a little, then into first with the easy, short-throw shift and off we sprint, no matter how hard we try not to… The 327 is a very easy car to drive. All the controls are well-balanced and the big car is surprisingly agile. Acceleration through the gears is quite remarkable, especially considering the reasonable fuel consumption. Declutch down to third (assuming you want to
go really fast), hit the gas and the Sting Ray's back squats, its nose rises, and cars fly by on the inside like they're standing still. A gorgeous V8 'wombah' accompanies the gear change, and the Sting Ray will cruise comfortably, depending on the gearing, at 3000rpm.

The Big Block is an altogether different animal. It's much the same to sit in, but the moment the engine starts its character changes dramatically. When a Small Block starts up, you know you're in a performance car, but when a Big Block starts up with that explosive roar, the pulse really starts racing.

The shift is still light, but the clutch is much heavier, and when you let up the pedal, the Sting Ray hustles forward like a desperate dog, hauling on its lead. The Big Block's road manners are very civilised so long as you only use the first centimeter of throttle pedal travel, and the car will pull away in top gear so gearchanging is seldom really necessary. For sensible driving just get it rolling in first and then shift into top and leave it there. For performance driving, shifting through the gears has an absolutely mindblowing effect: the wheels spin in every gear (depending on the tyres) the tail must be kept in line (but can be done so very controllably) and then the straight-line acceleration is collossal.

Believe you me, driving a 7-litre Sting Ray on the open road, with all those forces and fantastic noises, is the nearest thing to driving a CanAm car imaginable. And just like the owner told me, once you've driven a Big Block, nothing else will do. (So if you can't afford one, don't try one, and you'll never know what you're missing!) It may not be much faster than a well sorted 327 through the gears, but every second is precious to 100mph and nothing can compare with the sheer brutishness and massive torque of a Big Block, or its incredible fuel consumption: 5mpg is not unheard of if pressing on…

The bigger engine does not help the car's handling, and the 'Vette has more of a tendency to plough-on than with the smaller engined fitted, but in either Big or Small-Block form, the Sting Rays are very predictable to drive… Until you hit a poor road surface, or drive on the limit, or even just drive in the rain. Then their crude chassis design is only too obvious.

One final point. A well-sorted Big Block is one fine thing, a well-sorted Small Block is another fine thing. But a compromised or worn-out version of either will be a ghastly car, and far too many Sting Rays are just that. So don't be put off by the usual bad apples in the bunch. Just keep looking, and you'll find the one you want.

Buyer's Spot Check

Bodywork: One of the Sting Ray's strong points, the car is made from high-quality glass-fibre which seldom flexes or crazes. Occasionally star-burst crazing can be found on the bonnet where engine heat (especially on the Big Block cars) can cause expansion and contraction of the bonnet in winter. The areas around the tops of the wheelarch lips ❶ also occasionally show areas of stress, as do the body corners around the headlamp pods. Generally though, a Sting Ray's body should be in excellent condition; doors should not have dropped, panels should all fit. A poor paint job is a likely find – glass fibre is not easy to repaint – and after-market panels are frequently fitted. GM made the original Sting Ray panels from two metal dies. It was an expensive method but the results were quality panels. Check the inner surfaces of the car's panels – if they feel rough, they're after-market replacements.

Chassis: If after-market replacement panels are fitted to the car, special care should be taken when checking the chassis. Such panel replacement implies the car has been involved in an accident at some time, and accidents can crease chassis… the Sting Ray chassis did rust, especially to the points where

the rails sweep up over the rear axle ❷ where the trailing arms attach, and to the rear cross-member from which the spare wheel tray hangs ❸. Both areas are expensive to put right, and any Sting Ray with a rusty chassis should be left alone. Surface rust is usual and is nothing to worry about.

Engine and transmission: With both Small and Big Block Sting Rays, engines and transmissions should be trouble-free, except on high mileage cars where a routine full-rebuild can be expected at some time! All engines require frequent oil changes (every 3000 miles) with a very good oil, and few cars will have received such attention. Less frequent oil changes lead to camshaft wear. A poorly-maintained 427 is vulnerable, and these big engines are relatively delicate if not in perfect tune and maintenance.

Look for excessive smoke, listen for untoward noises, and expect to be truly startled by the performance if you haven't driven one before. If you're not, something is wrong with the engine. Remember solid-lifter engines will sound noisy to their top ends, and that fuel-injected 327s will only work properly if they are regularly used and flushed with fuel from time to time. Such 'fuelie' engines can be very expensive to put right as parts have to be specifically remanufactured. Small Block cars should run at 180°, Big Blocks at 200°.

Any healthy Sting Ray engine should hold 40lb oil pressure when hot and running, and pressure should not fall below 30lb at idle. Check the engine and engine bay carefully for originality, especially to the carburettor, air cleaner and rocker covers ❹. Transmissions are virtually indestructible and always create a certain amount of noise, but gearchanging should be easy and precise. If this is not the case the selector mechanism may need attention; it is situ-

A sight to intimidate anyone in a rear-view mirror – Hal Danby's 1963 Coupe is gorgeously aggressive. Above: The Coupe's snug and stylish interior. Right: Stomp on! Seven litres of sublime power corners flat

ated outside the gearbox casing and is therefore not difficult to attend to.

Suspension, steering and brakes: The condition of these areas can best be ascertained on the road. The Sting Ray should not wallow and should drive tightly in all departments. Worn bushes and suspension are common. The former can be expensive to put right, the latter is not. Expect the body to creak a little, and there is always scuttle shake on convertibles, but nothing too excessive. If you hear any clunking from the rear end be prepared for substantial outlay, either to replace the four universal joints on the drive-shafts ⑤ or possibly the differential itself. UJ wear is very common and not an easy job to put right. They should be lubricated regularly and seldom are. Steering will be generally a bit vague, but should have no play. Play will mean attention to either the four track-rod ends or the steering box: the former are cheap, the latter is not. The handbrake is seldom effective, but can be made to work properly. The drum braking system on pre-'65 cars is always pretty poor but can be made to work better with expensive modifications. The disc-braked cars should stop very well, but if they have been left standing for a long time or have not had their fluid regularly changed the cast steel calipers may have seized ⑥. Check to see if replacement stainless steel items have been fitted. If they haven't, budget for them.

Inside: Sting Ray interiors do wear but all parts are readily available from the profuse number of spare suppliers in America. Check the door trim panels for cracks, plus seat covers and headlinings. See also that the dashboard details are in period with the car, and check that the chassis, engine numbers and so on tally as they should. These plates are located under the glove box area ⑦.

CONTEMPORARY ROAD TESTS

In April '63 *Car and Driver* got their hands on a couple of coupés. They headlined, 'At long last America has a formidable weapon to challenge Europe's fastest grand touring cars on their home ground.' They went on to say, 'The key to the personality of the Sting Ray lies neither in the power available nor in the revised styling, but in the chassis. Up to now Corvette has been struggling to rise above a large number of stock components ... now the new independent suspension has completely transformed the Corvette but it still has some faults.'

Faults that were better noted by *Motor Trend* at the time, who more honestly reported on the Sting Ray's most obvious weakness, 'For a car that sells in the $4500-6000 range, it doesn't reflect the degree of quality we feel it should.'

C&D continued, 'A rigid front anti-roll bar in combination with a relatively stiff transverse leaf-spring in the rear reduces the resilience and independence of the suspension of each wheel with the result that even on mildly rough surfaces the car does not feel perfectly stable. On bumpy turns it's at its worst, veering from one course to another... but on a smooth course it comes in incredibly close to perfection.'

Autosport was one of the English magazines to test the Corvette and their tester remarked on the brakes of their fuel-injected '63 roadster, 'I have had the misfortune to drive cars with few brakes, but this car was the worst in this respect that I have ever experienced.'

There was much praise for the Sting Ray from all corners of the motoring arena, though, not least on the performance front. *Autocar* observed that only one car in its experience had accelerated faster than the Sting Ray, and that was a Zagato-bodied Aston Martin DB4!

The steering was generally considered rather vague but pleasant to use. *Car and Driver* declared in 1963 that 'there can be no conceivable need for power-assistance', although such an option might be more desirable today if wider tyres are used.

Of the 327, *Motor* and *Autocar*'s summing up remarks are worth noting. *Motor* declared: 'In most respects the Chevrolet Corvette Sting Ray is the equal of any GT car to be found on either side of the Atlantic. It falls down on refinement and wet road behaviour.' *Autocar* were more attacking. They concluded: 'The performance is certainly vivid, but if more attention had been given to such items as tyre adhesion in the wet and wipers that clear the screen effectively over 80mph, instead of some gimmicks which only attempt to give an impression of engineering thought, the Sting Ray would be a better car.'

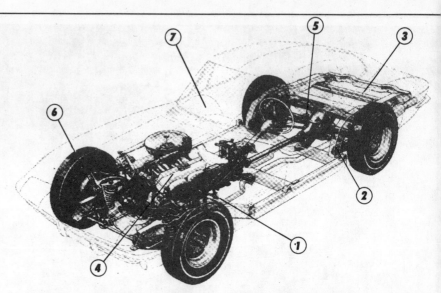

As for the Big-Block 427, I doubt whether more adulatory adjectives have ever been lavished on a mass-produced automobile. It was notably quicker than its 396cu in predecessor, and *Car and Drive* declared: 'Unless you're unlucky enough to encounter a Cobra 427, it's the wildest, hottest set-up going. With the normal 3.36 rear axle ratio it'll turn a quarter-mile that'll give a GTO morning sickness and still run a top speed of around 150mph.' But then *Car and Driver* made a very significant remark: 'The 327-engined version is still our favourite, but if you must go faster than anybody else... this is a pretty wild way to go.'

SPECIFICATION	1965 CHEVROLET CORVETTE STING RAY 327 (L79)	1966 CHEVROLET CORVETTE STING RAY 427 (L72)
Engine	90° V8	90° V8
Bore/stroke	4.00 × 3.25	4.25 × 3.76
Capacity	5359cc	7000cc
Valves	Pushrod ohv	Pushrod ohv
Compression ratio	11:1	11:1
Power	350bhp at 5500rpm	425bhp at 5600rpm
Torque	360lb ft at 3600rpm	460lb ft at 4000rpm
Carburettors	Single 4-barrel Holley	Single 4-barrel Holley
Transmission	4-speed Muncie	4-speed Muncie
Brakes	discs/discs	discs/discs
Front suspension	Ind by unequal-length wishbones, coil springs, anti-sway bar	Ind by unequal-length wishbones, coil springs, anti-sway bar
Rear suspension	Ind via drive-shafts and lower lateral links, transverse leaf, trailing arms	Ind via drive-shafts and lower lateral links, transverse leaf, trailing arms
Steering	Recirculating ball	Recirculating ball
Body/chassis	Glass-fibre/steel box-framed	Glass-fibre/steel box-framed
DIMENSIONS		
Length	14ft 7½in	14ft 7½in
Width	5ft 9in	5ft 9in
Height	3ft 11¾in	3ft 11¾in
Wheelbase	8ft 2in	8ft 2in
Weight	3130lb	3160lb
PERFORMANCE		
Max speed	130mph	152.5mph
0-60mph	6.2sec	5.4sec
Standing ¼ mile	14.9sec	12.8sec

PERFORMANCE COMPARISON (1967)

	Sting Ray 427	AC Cobra 427	Jaguar E-type 4.2
Max speed	142mph	165mph	119mph
0-60mph	4.7sec	4.2sec	6.7sec
Standing ¼ mile	13.6sec	12.4sec	15.3sec
Fuel consumption	9-13mpg	11-15mpg	16-19mpg

Production History

1963: The first year of the Sting Ray. Offered as coupé and convertible with one 327cu in engine in four different states of tune (250hp; 300hp-L75; 340hp-L76; 360hp-L84). Manual and automatic gearbox options, the latter only with 250hp and 300hp cars. 1963 Coupés are identifiable by their split rear window. Instruments are unique to the year, as are stainless steel details to the doors.

1964: 327cu in engine only and in four states of tune (250hp; 300hp-L75; 365hp-L76; 375hp-L84). Automatic available on 250hp and 300hp cars only, air conditioning on all but 375hp cars. Styling virtually identical, although single-piece wraparound screen on Coupé. Hood loses rectangular trim panels. Slight alterations to instruments, seat design.

1965: 327cu in and 396cu in engines now offered, the former in five states of tune (as for 1964 cars but also with 350hp-L79 specification), the latter in one state of tune (425hp-L78). Automatic and manual options as with 1964 cars. 350hp car available as manual only. Air conditioning availability as for 1964, and not available on Big Block cars. Disc brakes for all four wheels are made available this year and the vast majority of cars are fitted with them (drums were optional).

1966: 327cu in and 427cu in engines now offered, the former in just two states of tune (300hp-L79 and 350hp-L79), the latter also in two states of tune (390hp-L36 and 425hp-L72). Automatic available on 300hp 327 only. No air conditioning on 427s. Few styling changes – a new grille and Corvette script on the hood most obvious. No more roof vents for the coupé (functional and fake). Big Block cars have special hood.

1967: 327cu in and 427cu in engines available, the former as for 1966 cars with same transmission and air conditioning options, the latter in four states of tune (390hp-L36; 400hp-L68; 435hp-L71; 430hp-over 500hp-L88) with automatic option. 435hp car could have aluminium heads (L-89). 400hp and 435hp cars had three, two-barrel carburettors (termed Tri-power). 430hp, car had 12.5 compression and although claimed to output 430hp in reality 500hp was more likely. Less styling decor for 1967. New hood style for Big Block.

Engine Notes

● L84 engines were fitted with Rochester mechanical fuel injection.
● Aside from the L84s, most 327s employed a four-barrel Holley carburettor. Only the 300hp and 340hp cars did not, using a four-barrel Carter unit instead.
● All Big Block cars used a four-barrel Holley carburettor, except the aforementioned Tri-power cars.
● Although the barrel configurations may be the same, several different kinds of carburettors were used within these specifications. Check reference books for specifics.
● Compression for Sting Rays accounted for many of the output variations. The 300bhp 327s ran 10.5:1; the 350bhp ran 11.0:1; the other Small Blocks all ran 11.25:1. Of the Big Blocks, the L36 and the L68 ran 10.25:1, the L88 ran 12.5:1, the others all ran 11.0:1.

Production Figures

Year	Convertible	Coupé
1963	10,919	10,594
1964	13,925	8304
1965	15,376	8186
1966	17,762	9958
1967	14,436	8504

Prices

Sting Ray prices are very difficult to determine accurately at the best of times, and particularly at the moment. There are several reasons for this: firstly, there are so many different models with different specifications that it becomes almost impossible to distinguish between those which are marginally more desirable and those which are less so – with different engines, different gearboxes, different body styles, different option packs, different degrees of originality, different opinions as to which years are collectable and which factors are most important to the collector.

Secondly, those things which determine American prices are not always the things which affect European prices. The fitment of air conditioning, for example, is very desirable to collectors in America, but less so to we Europeans. Neither do we care so much about whether our Corvettes' engines are original to the cars – in the US they are obsessed with it. Also in America, the price variation between Big Block and Small Block cars tends to be wider, although, as most Sting Rays in this country are being imported from the States, this situation is rapidly changing.

Thirdly, Sting Rays are appreciating at such a rate in America that it is impossible to know which advertisers are underselling their cars, which are being reasonable, and which are just trying their luck.

Prices have virtually doubled for Sting Rays in America over the past year and are still rising fast, but the guide below is, we believe, pretty accurate at the time of going to press. The lower figure quoted is for a sound, useable car but which needs work or doesn't have an original engine. The higher figure is for an excellent all-round car, with its original engine, but without a full service history or other exceptional provenance. Concours show cars will be more expensive than the higher prices quoted. Fuel-injected 327s will be approximately £5000 more expensive than the prices quoted.

Year	Car	Price
1963	327 Coupé	£22,000-35,000
	327 Roadster	£17,000-25,000
1964	327 Coupé	£12,000-23,000
	327 Roadster	£15,000-24,000
1965	327 Coupé	£16,000-28,000
	327 Roadster	£16,000-30,000
	396 Coupé	£19,000-36,000
	396 Roadster	£22,000-40,000
1966	327 Coupé	£13,000-25,000
	327 Roadster	£14,000-27,000
	427 Coupé	£16,000-30,000
	427 Roadster	£17,000-35,000
1967	327 Coupé	£14,000-27,000
	327 Roadster	£15,000-30,000
	327 Roadster	£15,000-30,000
	427 Coupé	£18,000-60,000
	427 Roadster	£20,000-70,000
	L88 (any kind)	£150,000 plus

There are some price trends to note, and which are reflected in the figures above: with the exception of the split-window coupé, 1963 and '64 cars are worth substantially less than their later equivalents (due to their dreadful brakes); 1966 cars are worth less than their 1965 and '67 equivalents because they were neither the first of the disc brake cars nor the last of the line, and should therefore be considered relative bargains; a big premium is paid for 1967 cars, especially Big Blocks, where excellent examples are worth proportionally much more than any other Sting Ray model. Remember too that the top prices listed for 1967 cars apply to Tri-power cars and so cannot really be compared directly with 1966 equivalents. Notice too how restoration makes more economic sense on '67 cars than any others, as there is far more of a price margin between cheap and ultimate examples.

Spares Prices

Here's the good news – very good news: everything is available for Sting Rays, most of it cheaply, although some of it at a considerable price. It is possible, via spares suppliers and after-market parts manufacturers in the USA, to build any Sting Ray from scratch, right down to the chassis plates and air cleaner stickers. Aside from the rare performance packages, fuel-injection systems or the L88/L89 cars, most parts are very reasonably priced, although quality can vary enormously. Listed below is a selection of often-required spares and costs.

Engine rebuild (standard)	small block £1500
	Big block £2200
Exhaust system	£200 supplied
	£250 fitted
Clutch	£120 supplied
	£200 fitted
Universal joints (drive shafts)	£120 supplied
	£270 fitted
Full re-bushing	£250 (front) fitted
	£200 (rear) fitted
Radiator	£120 supplied
	£150 fitted
Headlamp motors	£300 supplied
Carpets and seat covers	£650 supplied
Stainless steel brake calipers	£280 supplied
	£330 fitted

You can get all the bits... but get them right!

Specialists

BC Classic American Automobiles Ltd, 9 Oakleigh Gardens, Edgware, Middx HA8 8EA (tel: 01-958 1756). Probably Britain's largest importers of Sting Rays, Managing Director Bernie Chodash has been a specialist for many years and his prices are very reasonable.
St Albans Sports Cars, 52 Great Suffolk Street, London SE1 (tel: 0727 47778 or 0836 291969). The man responsible for preparing virtually all the Chevrolets currently completing in HSCC rounds. Chris Allen is strongly recommended for his mechanical knowledge. He also buys and sells Corvettes at sensible prices.
Claremont Corvette, Corvette Centre, Snodland, Kent ME6 5NA (tel: 0634 244444). Europe's largest stockists of secondhand (and new) Corvette parts and associated automobilia. Long-established operation run by author Thomas Falconer. He usually has six cars for sale at a time too.
Corvette Kingdom, Unit 4, High Street, Stalham, Norfolk (tel: 0692 81374). Full facilities for Corvette restoration and repair in East Anglia.
South-East Corvette Centre, 416 Montagu Road, Edmonton, London N9 0ET (tel: 01-803 8311). North London's long-time Corvette specialists who will undertake any Sting Ray work. Owner Peter Hollingworth also buys and sells the occasional Sting Ray.

CLUB

Classic Corvette Club UK: Celebrating its tenth anniversary this year, the club now numbers 289 members. Membership costs £15 pa, and benefits include receipt of their quarterly magazine and newsletters too. Annual Open Day is on September 17 at Beaulieu. Club details from Geoff York, Poor Parsonage Close, Bishop Tachbrook, Warks (tel: 0926 81140).

BOOKS

There are enormous numbers of Corvette books on the 1963-67 cars. Listed below are a personal selection of some of the best.
Corvette. America's Star-spangled Sports Car, by Karl Ludvigsen (£29.95). Heavy going, definitive volume on the model. Out of print.
Early Chevrolet Corvettes, 1953-67, by Thomas Falconer (£7.95). Brief introduction with lots of pictures. Useful statistics.
Corvette Black Book, by Michael Antonick ($9.95, plus postage). Essential compact Corvette note book detailing all numbers and annual changes. Very useful for the buyer, but only available in the USA or by post from Michael Bruce Associates Inc, PO Box 396, Powell, Ohio 43065.
Corvette Sting Ray, 1963-67, Super Profile, by Bob Ackerson (£5.95). Usual Super Profile format – rather thin but easy to read and useful introduction for potential car buyer. Certainly good value.
Car and Driver on Corvette, 1956-67 and *Road and Track on Corvette, 1953-67* (£6.95 each). Brooklands Books.
Corvette Restoration, State of the Art, by Michael Antonick (£12.95). The story of a '65 Big Block rebuild. Incredibly detailed. Invaluable.
Corvette. 1953 to present, by Richard Nichols (£9.95). Large format, lots of illustration, quite comprehensive, excellent value.
The Secrets of Corvette Detailing, by Michael Antonick (£9.95). Essential guide to getting it right.
Maximising Your Corvette's Potential, by Michael Antonick (£8.95). Useful guide to restoration and race preparation.
The Complete Corvette Restoration Guide, by Noland Adams (£49.95). Very expensive but essential guide for the restorer. Ultimate reference book and well put together.
The Best of Corvette News, (£32.50). A compilation of the best articles from *Corvette News* over the years. Very interesting period stuff, but a bit of an indulgence at the price.
Corvette! 30 Years of Great Advertising, (£15.95). Tells you nothing, but lovely to look at. Every owner should have a copy.
Vette Vues Fact Book, 1963-67, by M.F. Dobbins (£19.95). Absolutely everything you need to know to ascertain the originality of your Sting Ray, and a lot more besides. Strongly recommended.

ACKNOWLEDGEMENTS

Our grateful thanks are due to Nick Harper for the generous loan of his 1966 427 roadster and to Hal Danby for bringing his 1963 Coupé along to our photo session. Bernie Chodash arranged for the 427's loan and helped with pricing and buying information. Chris Allen gave invaluable buying and restoration advice, and bought Hal's split-roof 327 to Suffolk for testing. Thanks too for the PR staff at USAF Lakenheath for their co-operation.

OWNER'S VIEW

Hal Danby only ever wanted a split-window Sting Ray, and he finally bought one before he moved back home from America. Then he took it motor racing…

There was a marvellous Scalextric kit of a Corvette Sting Ray that I vividly remember from when I was a teenager. I made several of them at the time, and I suppose ever since then I'd always wanted a real one. And it had to be a split-window coupé too, because that was what the Scalex' car was.

I was able to buy one when I was working in America in 1983 – the prices were low at that time and it was scarcely an expensive indulgence. I had started working in America in 1978 on the West Coast and had just sold my business, but before returning to England had to stay in the US for a couple more years – I bought the Corvette to help pass the time, and then brought it back to Britain with me.

The Sting Ray was in lovely original condition and had only had one previous owner. She was a TWA air stewardess who had been given the car by her father for her 21st birthday. He even found the license number TWA 727 (which I still have with the car) but then only let her drive it after she could recite the owner's manual word for word! Subsequently she became something of a Corvette expert and maintained it superbly. She sold it only because she was moving to New York and felt that it was a Californian car that would soon be home-sick in the east. Little did she know I'd be bringing it here!

When I bought it to Britain it was easier to maintain in good condition. It's amazing what a few English winters can do to a car… In California it had never been in the rain in 20 years, whereas now there is rust on the suspension parts and so on. The original owner would be horrified… And then I decided to take it motor racing with the HSCC.

I started racing when I was 19, I suppose, and have always loved it. I used to campaign a Mk1 Escort which had a Martin Formula 1 V8 engine in it, and a few other things, but when I was in America I missed English club racing dearly. On my return the Corvette's destiny was set. In some respects it seemed a bit sacrilegious to modify such an exceptionally original car for racing but I've kept all the original bits and nothing more than a different carburettor was needed to make it competitive against all the European jobs.

It's a very easy car to race, and it's a very easy car to live with. Everything is so robust on it – good, solid American engineering. It could probably be half the weight, but that's not really the point. Everything still works on that car, nothing has fallen off or broken. It's also very easy to repair – although there are certain jobs I don't do myself, but that is more because over the years I've lost enthusiasm for getting grubby, rather than anything else. I adore driving it, but I must admit I don't like being the centre of attention the whole time, and that is a problem with the Sting Ray. People are always looking at you, and that includes vandals. It's not very private.

The obvious comparison to make with the Sting Ray is the Jaguar E-type, and I've had one of those and was looking at them again last year before the prices went shooting up. Alas I missed the boat in the end, but there is more room in the Corvette. Having said that, I had to put a smaller steering wheel on the car so I could fit in it properly, and now I can't see the instruments. It also mists up like crazy; there's an annoying gear rattle that I just can't get rid of; the seats are horrible (I can't really get comfortable) and then there's the worst aspect of the early Sting Rays – those terrible brakes. Chris Allen has improved mine and they now work very well, but in original order were awful, really foul. The drums are much too small and when they lock up, you don't know which wheel will go first. And when racing in the wet that is very nasty.

It's really quite a couth car – it doesn't rattle and it certainly doesn't smell of hot glass-fibre, and thanks to the excellent cool air ventilation it doesn't get too hot inside either. The engine is absolutely wonderful. I just love it. It is so flexible and you can have all sorts of boy racer fun with it. You can put your foot down on tight bends and the back will come right round. It's generally very predictable and easy to drive, but I have been scared of it a couple of times. There is something that tells me from time to time not to fool with it. I raced Roy Jordan's 1964 coupé last year, which is a little more modified to the suspension than mine, and I had felt much safer in it.

As a daily car the Sting Ray is entirely practical. Only fuel consumption detracts, but although when racing I get little more than 5mpg, on the road that rises to about 15mpg, which I don't think is at all bad for all that power.

I think it is the finest sports car I have ever owned, although it compares well with my old E-type two-seater. But the Corvette is much rarer in this country, and it's not as valuable either. One day they will be worth a lot, but in the meantime we'll do a bit more racing and have a lot more fun.